Professional Development for Teachers of Mathematics

1994 Yearbook

Douglas B. Aichele
1994 Yearbook Editor
Oklahoma State University

Arthur F. Coxford
General Yearbook Editor
University of Michigan

**National Council of
Teachers of Mathematics**

ISSN 0077-4103
ISBN 0-87353-366-6

Printed in the United States of America

Contents

iii

Part 2: Initial Preparation of Teachers of Mathematics

Part 3: Professional Development for Practicing Teachers of Mathematics

Preface

The publication of the NCTM *Curriculum and Evaluation Standards for School Mathematics* and the *Professional Standards for Teaching Mathematics* has provided a new vision for the school mathematics curriculum and for teaching school mathematics effectively. These documents have created the basis for programs of professional development unparalleled in the history of the teaching of school mathematics. Jointly, they support professional development programs that will help both preservice and in-service professionals meet the Standards for curriculum, evaluation, *and* teaching.

This 1994 NCTM Yearbook has been designed to provide a stimulating collection of articles on the professional development of teachers of mathematics. We sought contributions that addressed general issues and perspectives related to professional development; the initial formal preparation of teachers of mathematics at the elementary, middle and junior high, and senior high school levels; and the continuing growth of practicing teachers of mathematics. More than 100 proposals were submitted to the editorial panel for review, and this final publication includes twenty-nine contributions from sixty-one authors.

The yearbook is organized into three major parts. Part 1 sets the stage for the book by examining the issues associated with professional development. Based on known research, these articles address the stages of the continuum of professional development from general to mathematics-specific perspectives. The articles in Part 2 address the initial preparation of teachers of mathematics in grades K–12. A call for changing preservice programs is followed by descriptions of exemplary mathematics teacher education programs at all levels, K–12, including a model for a preservice program to prepare elementary mathematics specialists. Part 3 includes articles that focus on professional development for practitioners. These articles describe very creative projects and programs from the educational and business communities. High levels of cooperation and collaboration among those constituencies committed to reforming mathematics education are showcased throughout. A common theme that emerged among participating practitioners was that "becoming a learner again makes them better teachers."

The 1994 Yearbook reflects the untiring efforts and commitments of many people. The editorial panel developed guidelines, reviewed and selected manuscripts, made valuable suggestions for improving manuscripts, and shaped the yearbook's overall direction. The panel included the general editor, Arthur F. Coxford, and four individuals with great insight into the professional development of teachers of mathematics. My sincere thanks and appreciation are given to these individuals:

Margaret S. Butler	Bartlesville Public Schools, Bartlesville, Oklahoma
Arthur F. Coxford	University of Michigan, Ann Arbor, Michigan
Peggy A. House	Northern Michigan University, Marquette, Michigan
William A. Juraschek	University of Colorado—Denver, Denver, Colorado
Glenda Lappan	Michigan State University, East Lansing, Michigan

Throughout the yearbook production process, from planning meetings to the final editing of manuscripts and preparation for printing, the editors and panel have been supported well by the outstanding NCTM headquarters staff. Thank you.

DOUGLAS B. AICHELE
1994 Yearbook Editor

1

Professional Development and Teacher Autonomy

Part 1

Kathryn Castle
Douglas B. Aichele

T O UNDERSTAND the highly complex issue of professional development for teachers of mathematics, one must address such questions as the following:

- How did we get where we are today?
- What forces, both internal and external to the mathematics education community, influence the professional development of teachers of mathematics?
- How can these forces be redirected so that they are facilitating and focused for good?

The evolution of mathematics education in the United States has been highlighted by research, reports, recommendations, and contributions from special commissions, agencies, and professional organizations concerned with improving the learning and teaching of mathematics. Mathematical education for an enlightened citizenry has always been a priority of U.S. society. Because of this importance to society in general, forces external to the mathematics education community have directed and contributed to the changes being experienced today.

Mathematics education is largely an effect of the past, especially the recent past since 1960; the future will undoubtedly be influenced by the issues and concerns of the present (Aichele 1978). Though the early 1950s can be characterized as a time of emerging activity in mathematics education, the period from 1957 to the mid-1970s is generally regarded as the era of "modern math" or "new math." It was during this period that a revolution was literally taking place in the school mathematics curriculum. The changes that were appearing in the content of school mathematics at this time were so far-reaching and profound that they appropriately characterized a revolution that was caused principally by advances in mathematical research, automation, and computers (National Council of

Teachers of Mathematics [NCTM] 1961). This revolution was also fueled by wide-spread public dissatisfaction with the mathematical preparation of entering college students and the fear that the Soviet Union would gain supremacy in space. The single event that was most significant in forming this public attitude was the launching of Sputnik in 1957. The emphasis and thrust of the programs were clear during this period: to produce quality mathematics curricula for the college-bound student. This would soon change, however, since attention would be given to developing programs for less-able students and to expanding educational opportunities for all youth (Aichele 1988).

The revolution in the content of school mathematics was accompanied by a critical shortage of teachers of mathematics. Professional development for teachers of mathematics as we know it today was in its infancy. Teacher-preparation programs could be characterized as "content driven" and teacher in-service programs as "quick fix" remedies. As agents responsible for carrying out a fragmented national agenda to raise the level of student performance, teachers had no time to collaborate with colleagues, plan for instruction, or reflect on their practices. They found themselves in the unenviable position of literally trying to "learn enough math to get ready for the next class."

The reaction to the "new math" era was found in the "back to the basics" movement of the late 1970s. Though broader than mathematics, this movement (applied to mathematics) prescribed a very limited mathematics curriculum and focused almost exclusively on computation. Some of the issues that accompanied this movement were tests of minimum competency for entering employment, school and teaching accountability by achievement testing, and pedagogical theories based on behavioral objectives. Assumptions underlying these programs and their subsequent implementation were in direct contrast to those underlying the "new math" movement and its implementation.

In 1975, an organized response from the mathematics education community was provided through the National Advisory Committee on Mathematical Education (NACOME) Report. In its overview and analysis of U.S. school-level mathematics education, it reported that "the principal thrust of change in school mathematics remains fundamentally sound, though actual impact has been modest relative to expectations" (Conference Board of the Mathematical Sciences 1975, p. 21). The recommendations of the NACOME Report have been viewed by many as the catalyst for defusing the commonly misunderstood effects of the "new math" era and for providing the direction for the future. The recommendations related to teacher education (professional development) recognized the growing needs of teachers in a changing mathematics environment.

Since the beginning of the decade of the 1980s, leadership from the professional mathematics associations has been continuous and directed at all segments of the profession and society. This leadership began with the publication of the *Agenda for Action* (NCTM 1980), a document advising society of the directions of mathematics education during the decade. The *Recommendations on the Mathematical*

Preparation of Teachers (Mathematical Association of America 1983) and its sequel, the *Guidelines for the Continuing Mathematical Education of Teachers* (Mathematical Association of America 1988), represented efforts by the mathematics community to address the critical shortage of adequately qualified teachers of mathematics by suggesting types of programs, ways of meeting the needs of teachers, and general mathematical content. Building on the momentum of the 1980s, standards were subsequently developed for the school mathematics curriculum, teaching, and evaluation (assessment). The *Curriculum and Evaluation Standards for School Mathematics* (NCTM 1989) and the *Professional Standards for Teaching Mathematics* (NCTM 1991) have provided the mathematics education community with the needed vision for mathematics learning and teaching in the twenty-first century. These nationally recognized standards have been adopted by the National Council for the Accreditation of Teacher Education for use in accreditation and have become the model for standards created by other professional organizations. The *Professional Standards for Teaching Mathematics* (NCTM 1991) is unique in the evolution of mathe-matics education in that its standards built an irrefutable link between professional development and curriculum.

In addition to the teaching standards held by the profession, numerous external forces are being exerted on teachers today that influence and direct their professional development. For example, accreditation agencies, local, state, and national initiatives, induction programs, funding agencies, career ladders, and merit pay greatly influence teachers' beliefs, attitudes, expectations, and classroom practices. However, to be effective *professional* practitioners, teachers must develop their own vision for professional development. This requires a level of teacher autonomy reflecting an internal drive toward professionalism.

MOVING TOWARD PROFESSIONAL AUTONOMY

The professional development of mathematics teachers is a highly complex issue that has the potential to transform the entire field of education. Mathematics teachers develop professionally in the same ways all other teachers do but with a specific focus of applying professional knowledge within a meaningful and relevant mathematical context for the improvement of the mathematical understanding of children and youth.

Professional development takes many forms, but true professional development, in the sense of resulting in meaningful and long-lasting qualitative change in a teacher's thinking and approaches to educating, is an autonomous activity chosen by a teacher in search of better ways of knowing and teaching mathematics. Professional development activities that are externally mandated or coerced by a power hierarchy are false because they do not result in development as a qualitative change. Externally imposed professional development activities, although well intentioned, are doomed to failure, like other passing educational fads on the junk heap of discarded simplistic solutions to complex problems. In short, one cannot "empower" another professionally. The term *empowerment* implies an external agent transferring "power" or knowledge to a teacher. Professional knowledge cannot be transferred. Rather, it is constructed by each individual teacher bringing his or her "lived experiences" as a learner and teacher to an educational setting and interacting with the environment in a way that relates new knowledge to previously constructed knowledge in an attempt to make the best sense of the new knowledge. Professional knowledge actively constructed by teachers allows teachers to function as autonomous decision makers who act in the best interests of students regardless of any externally imposed reward system. Autonomous teachers act on their best professional knowledge, whereas heteronomous teachers merely attempt to carry out the mandates of others in authority positions.

THE MEANING OF PROFESSIONAL AUTONOMY

Teachers move toward professional autonomy as they continue to construct their ideas about mathematics and how it can best be taught to others. Professional autonomy involves the ability to exercise good judgment. Autonomous teachers make decisions after carefully considering relevant variables, including the effects their decisions have on others. Autonomy is self-regulation, the ability to decide for oneself without having to be told by others. Autonomous teachers are capable of distinguishing between appropriate and inappropriate actions on the basis of an internally constructed standard for behavior (Kamii 1992). They make decisions about appropriate practice on the basis of their previously constructed knowledge of children and youth and how they learn best. They do not abdicate their decision-making responsibilities to other teachers, principals, or text and test publishers. For example, autonomous teachers are likely to read and critique professional guidelines such as the NCTM *Standards* in light of what they know good practice to be. Autonomous teachers are more likely to be able to continually improve and revise professional standards than to merely carry them out. Autonomy leads to continued construction and reconstruction of knowledge, which amounts to advanced progress in the field. Autonomy as an aim for education requires creating conditions for teachers to act autonomously (Wilucki 1990).

EXPERIENCES THAT FOSTER PROFESSIONAL AUTONOMY

Both preservice and in-service teacher education programs can foster professional autonomy by appreciating the unique lived experiences of teachers in the programs and by providing opportunities for teachers to experience ownership of the programs. Teachers can have a voice in the programs by sharing their own lived experiences and teacher stories with others (O'Loughlin 1992). Through reflecting on lived experiences and having dialogue with others, teachers begin to reconstruct what it means to be a learner and teacher of mathematics.

Professional autonomy is fostered through making choices and decisions, setting goals, reflecting on teaching practices, exchanging points of view with others, and engaging in dialogue with others on relevant educational issues. Programs that foster professional autonomy put individuals in control of their own learning, including the selection of goals and means of assessment.

Choice, Decision Making, and Goal Setting

Having choice in an educational program equates to personal freedom. When individuals are free to choose, they are more likely to think about the choices and decide what is the best choice among the alternatives. When teachers are told what to do, they do not think; they just respond. Since the activity was not of their choice, they do not find it personally meaningful—that is, it does not connect with their previous knowledge in a way that makes sense to them. Individuals who do not experience choices in teacher education programs are more likely to view theory, research, and practice as fragmented and disconnected. Individuals who actively choose are more likely to see connections and engage in practice that reflects a cohesive view of the relationship of teaching to learning. To the extent that teachers are told what to teach and how to teach, they are robbed of the opportunity to choose for themselves and ultimately to provide the opportunity of choice to children and youth.

Having choice also involves choosing personal goals, including ways to determine whether goals have been met. Goals that are chosen represent program ownership and are more likely to be achieved because they have personal relevance.

Reflective Practice

Through reflecting on teaching practices, teachers develop autonomy. Reflective practice engages teachers in thinking about good teaching and reflecting on the meaning teaching has for students. An example of reflective practice is journal writing in which teachers write about their lived teaching experiences. Writing and reflecting on experiences helps teachers develop insights into their own behaviors and how they affect students through their teaching practices. Reflective insights provide a deeper and richer understanding of what it means to teach, thus contributing to professional knowledge used to make autonomous decisions.

Exchanging Points of View

Professional autonomy is developed when teachers have opportunities to share their views with others and to hear and debate the views of others. Discussions, debates, and seminars give opportunities for sharing the views of others. Through sharing views, teachers become better at expressing and clarifying their reasoning on issues. When teachers are asked questions about their ideas and are asked to explain them, they reach new depths of understanding of issues. As they consider alternative viewpoints on an issue, they begin to think about the issue in new and more insightful ways.

Through exchanging points of view, teachers develop an appreciation for diversity of thought. They become better at seeing another's perspective, which leads to better decision making. Considering an issue from multiple viewpoints helps in deciding what is best for the majority of those involved.

Exchanging points of view can occur in group discussions of case studies. Teachers can debate the most pressing issues in these case studies. Such problem-solving discussions help teachers decide on the best ways to handle teaching problems. Teachers can be encouraged to write and present their own case studies or situations where they have experienced a difficult problem. Writing personal narratives and sharing them with others helps in appreciating the complexity of teaching and the individuality of each teacher.

Another example of exchanging viewpoints would be participation in a collaborative project. Group membership allows for discussion and debate within a secure environment in which one feels acceptance, "belongingness," and a sense of personal contribution to the group. Through such participation, teachers form support networks for one another and work cooperatively on mutual goals, tasks, and projects.

ENVIRONMENTS THAT FOSTER PROFESSIONAL AUTONOMY

Teacher education programs and schools that foster autonomy also encourage personal responsibility for learning and decision making. Such environments provide opportunities for shared decision making, working collaboratively with others, posing problems to tackle, determining criteria to apply to one's work, engaging in self-assessment, and setting goals (Boud 1988). In implementing a teacher education program designed to promote autonomy, Fosnot (1989, p. 21) writes:

> Rather than dispensing a list of prescribed methods of instruction to preservice teachers for their use, these teacher candidates themselves need to be immersed in an environment where they are engaged in questioning, hypothesizing, investigating, imagining, and debating. They need to be part of a community that actively works with them as *learners* and then allows the experience to be dissected, evaluated, and reflected upon in order for principles of pedagogy and action to be constructed.

Fosnot engaged in this type of interaction with students and charted the transformation process that occurred in movement toward autonomous teaching.

In an interview by Brandt (1991), Goodlad says that placing student-teachers in school environments that promote the cloning of cooperating teachers should be avoided because it perpetuates present practices and does not encourage improvement beyond traditional ways of teaching. Student-teachers develop autonomy in teaching through being encouraged to plan their own objectives, units, themes, and projects under the guidance (not control) of a master teacher.

Teacher education programs that help student-teachers reflect on their goals and communicate them to cooperating teachers will prevent the mindless replication of inappropriate practices imitated by student-teachers who lack such a foundation (Castle and Rahhal 1992). Negotiations of what preservice teachers do during the student-teaching period occur very early in the experience (during the first two weeks) and regulate the preservice teacher's planning and implementation of learning activities (Grant and Castle 1990). Student-teachers who are not able to negotiate their goals are more likely to merely carry out the traditional practices of their cooperating teachers (Castle and Meyer 1991).

AUTONOMOUS TEACHING

Autonomous teachers promote autonomy in children and youth. Autonomous teachers are self-directed learners who question, study, and search for answers from a need to know. They can articulate to others their views on education issues and construct their own theories of what constitutes good teaching. Through interactions with others, they clarify their ideas and reconstruct what it means to be a professional teacher. This process of reconstructing leads to new and better approaches to teaching. Autonomous teachers construct personally meaningful professional knowledge resistant to education fads or external mandates. They have constructed their professional knowledge, not borrowed it from experts, and are therefore more confident in what they know. They appreciate the constructive process and are better able to foster it in children and youth.

REFERENCES

Aichele, Douglas B. "Historical Overview." In *Mathematics Education in Secondary Schools and Two-Year Colleges,* edited by Paul J. Campbell and Louise S. Grinstein, pp. 3–22. New York: Garland Publishing, 1988.

———. "Mathematics Teacher Education: An Overview in Perspective." In *Mathematics Teacher Education: Critical Issues and Trends,* edited by Douglas B. Aichele, pp. 9–24. Washington, D.C.: National Education Association, 1978.

Boud, David. *Developing Student Autonomy in Learning.* 2nd ed. New York: Nichols Publishing Co., 1988.

Brandt, Ron. "On Teacher Education: A Conversation with John Goodlad." *Educational Leadership* 49 (1991): 11–13.

Castle, Kathryn, and Jane Meyer. "Student Teacher Autonomy: Negotiation of Student Teaching Experiences." *Journal of Early Childhood Teacher Education* 12 (1) (Winter 1991): 8.

Castle, Kathryn, and Kelly Rahhal. "Moving toward Developmentally Appropriate Practice in Primary Teaching." *Journal of Early Childhood Teacher Education* 13 (1) (Winter 1992): 3–6.

Conference Board of the Mathematical Sciences, National Advisory Committee on Mathematical Education. *Overview and Analysis of School Mathematics, Grades K–12.* Reston, Va.: National Council of Teachers of Mathematics, 1975.

Fosnot, Catherine. *Enquiring Teachers, Enquiring Learners: A Constructivist Approach for Teaching.* New York: Teachers College Press, 1989.

Grant, Kay, and Kathryn Castle. "Theory into Practice: Student Teachers' Construction of Knowledge." *Journal of Early Childhood Teacher Education* 11(1) (Winter 1990): 13–14.

Kamii, Constance. "Autonomy as the Aim of Constructivist Education: How Can It Be Fostered?" In *Project Construct a Curriculum Guide—Understanding the Possibilities,* edited by Deborah G. Murphy and Stacie G. Goffin. Jefferson City, Mo.: Missouri Department of Elementary and Secondary Education, 1992.

Mathematical Association of America. *Guidelines for the Continuing Mathematical Education of Teachers.* Washington, D.C.: The Association, 1988.

———. *Recommendations on the Mathematical Preparation of Teachers.* Washington, D.C.: The Association, 1983.

National Council of Teachers of Mathematics. *An Agenda for Action: Recommendations for School Mathematics of the 1980s.* Reston, Va.: The Council, 1980.

———. *Curriculum and Evaluation Standards for School Mathematics.* Reston, Va.: The Council, 1989.

———. *Professional Standards for Teaching Mathematics.* Reston, Va.: The Council, 1991.

———. *The Revolution in School Mathematics, a Report.* Washington, D.C.: The Council, 1961.

O'Loughlin, Michael. "Engaging Teachers in Emancipatory Knowledge Construction." *Journal of Teacher Education* 43 (1992): 336–46.

Wilucki, Belinda McCully. "Autonomy: The Goal for Classroom Teachers of the 1990s." *Childhood Education* 66 (1990): 279–80.

2

Teacher Education as an Exercise in Adaptation

Thomas J. Cooney

It was said of Captain Scott, a great explorer, that he was a strange mixture of the dreamy and the practical, and never more practical than immediately after he had been dreamy.

—William Barclay, *The Gospel of Luke*

I RECALL a *Peanuts* cartoon of many years ago in which Linus said with great frustration, "How do you expect me to do new math with an old math mind?" I would twist this question slightly and suggest that a teacher frustrated with the recent calls for reform and further demands placed on a schedule that would suffocate most mere mortals might cry out in anguish, "How can you expect me to teach new math with an old-math mind?" The answer seems obvious: You cannot. You cannot, that is, without the support of colleagues, administrators, parents, and, perhaps most important, a vision of teaching mathematics that suggests that life in the classroom can be different.

Our ability to realize a vision, that is, to change, is largely a function of our ability to be adaptive agents. To be sure, being an adaptive agent in the classroom requires a great deal of knowledge about mathematics, pedagogy, and the psychology of learning. But it also requires a certain orientation toward that knowledge. The lenses through which we see our world influence much of what we do. But these lenses are not created out of the abstract; they are context specific. Thus the issue of teacher education becomes immensely complex as we strive to honor what teachers bring to the enterprise while at the same time assist them in becoming adaptive agents in their classrooms.

THE NOTION OF ADAPTATION

Ultimately, any attempt to reform the teaching of mathematics is an exercise in adaptation from what we are *able to do* to what we *want to do*. The notion of

9

able to do involves teachers' knowledge of mathematics and of the teaching and learning of mathematics. In the absence of such knowledge, the process of adaptation is severely limited, if not impossible. But sometimes teacher education programs focus too much on correcting the deficiency in teachers' knowledge (i.e., giving them "more" of something) to the exclusion of considering possibilities of how that knowledge can transform the common, everyday activities of teaching.

Von Glasersfeld (1989) drew our attention to the importance of context in the creation of knowledge when he identified the following two principles of constructivism:

1. Knowledge is not passively received but actively built up by the cognizing subject.
2. The function of cognition is adaptive and serves the organization of the experiential world, not the discovery of ontological reality.

Everyone has had the experience of communicating with someone for whom the meanings of the words we had spoken were different from ours. The result is a lack of mutual understanding. I recall an incident in which a teacher became uneasy as the decibel level in her classroom steadily increased. With all the sternness the young and anxious teacher could muster, she proclaimed in a voice that riveted the second graders, "I demand pandemonium!" It worked. The medium had belied the message, but the students responded to what they perceived to be the message: They needed to get quiet. When people speak to us, we interpret what they are saying in the context of the network of ideas and experiences we have had. Knowledge, of any sort, is a product of those experiences. This statement is true for knowledge about teaching as well as for knowledge about mathematics. That knowledge about teaching may be tacit in no way diminishes its importance. Indeed, the very act of making the implicit explicit may be a first step in laying a foundation for change.

Von Glasersfeld's second principle reminds us that human behavior adapts to the perceived contexts in which we find ourselves. As Kuhn (1970) has so persuasively argued, knowledge structures are necessarily contextual. The implication for teacher education is that acquiring new methods of teaching mathematics is necessarily and fundamentally connected to our conception of what it means to teach mathematics. For the preservice teacher, this conception may be the result of accumulated experiences as a student of mathematics; for the in-service teacher, conceptions are more likely rooted in the practical—what seemed to work yesterday. In either instance, it makes no more sense to base our teacher education programs on the assumption that teachers are *tabula rasa* than to assume that students enter their classrooms void of a wide range of conceptions of mathematics.

It follows that what teachers internalize from their teacher education experiences is a function of their experiential world, not that of the teacher educator.

That the teacher educator may see conflict between current practice and reform as set forth in NCTM's *Curriculum and Evaluation Standards* (NCTM 1989) and *Professional Teaching Standards* (NCTM 1991) does not ensure that the teacher sees any such conflict. The importance of this realization is that it warns us about giving answers for which there were no questions. From this perspective, it seems clear that one or two courses or workshops will typically fail to provide the necessary intervention for changing the teaching of mathematics. Lappan et al. (1988) addressed the importance of continuity between a teacher education program and teachers who were trying to effect change in their classrooms. These teacher educators found that teachers could acquire considerable information in a single workshop but that the translation of this information into their classroom teaching required a sustained in-service program of at least two years' duration in which teachers were given intellectual and emotional support as well as technical assistance.

Accepting the premise that adaptation should be at the core of teacher education programs does not diminish the importance of acquiring "more," as mentioned earlier. But it requires the realization that "more" can contribute to change only if it is seen as a means for achieving change rather than as an end in itself.

UNDERSTANDING IN A TEACHER EDUCATION CONTEXT

Reviews of teacher education programs by Fitzsimmons (1991) and Brown, Cooney, and Jones (1990) revealed few articles in which authors explicitly discussed the epistemological foundations underlying their programs. Descriptions of programs emphasized goal-oriented outcomes, such as teachers' practicing new skills, using problem-solving strategies, or incorporating technology into their teaching. One senses that teacher education is inherently a practical matter—a notion reflected in the rather practical goals stated for many teacher education programs.

Still, research reveals some areas of concern in the practical world of teacher education. Although no firm evidence can be cited that knowledge of mathematics is related to effective teaching, it is difficult to imagine a rational argument that knowledge of mathematics should not be an integral component of a mathematics teacher education program. At the elementary level, such studies as those of Wheeler and Feghali (1983); Graeber, Tirosh, and Glover (1986); Fisher (1988); and Mayberry (1983) indicate that teachers' knowledge of mathematics is shallow and that this deficiency represents a real impediment to achieving reform. At the secondary level, virtually no research has been reported on teachers' knowledge of mathematics beyond defining mathematical knowledge in terms of courses taken at the college level. On the basis of this inadequate definition, no significant statistical or educational relationship appears to exist between knowledge of mathematics and teaching effectiveness (Begle 1968; Eisenberg 1977), yet surely adaptation cannot become a reality in the absence of a strong knowledge of mathematics.

There is evidence that teachers' *orientation* toward mathematics is limited. Case studies have revealed that some teachers, perhaps even many, treat mathematics as a cut-and-dried subject (Kesler 1985; McGalliard 1983). This orientation appears to be rooted in experiences that occur long before teachers begin their formal mathematics education courses (Meyerson 1978; Owens 1987; Helms 1989; Wilson 1992), and teachers seem to be resistant to change unless significant intervention through in-service programs is initiated (Bush 1983; Ball 1988).

These findings suggest that teachers may be lacking a sophisticated view of mathematics. Owens's (1987) study suggests that this orientation is not resolved and may be perpetuated by a study of advanced college mathematics, with its heavy emphasis on formalism and symbolism. Wittmann (1992), in making the case that mathematical formalism contributes to a "broadcast metaphor" of teaching, tells the following story about a primary-level preservice teacher who reacted to a proof of Euler's polyhedron theorem (p. 111):

> After the proof a student intervened and asked: "Was this really a proof?" I was struck, because this question is usually asked if a proof is evaluated as not [being formal enough] and with this group of students I had expected an appreciation of non-formal proof. However, when [I asked], "Why not?" I received a surprising and instructive answer: "Because I understood it."

Wittmann's main point is that the formalism of mathematics tends to impose an approach to the teaching of mathematics that casts mathematics as a "steep and harsh terrain of abstract language that separates the mathematical rain forest from the domain of ordinary human activity" (Steen 1988, p. 611).

Another area of concern is the means by which teachers assess their students' understanding of mathematics. In a survey of 201 secondary school mathematics teachers (grades 7–12) in the state of Georgia, Cooney (1992) found that 57 percent of the teachers identified computation or simple one-step problems as representative of items that assess a deep and thorough understanding of mathematics. The following items were typical of those thought to assess a deep and thorough understanding of mathematics:

- $4\ 1/3 + 2\ 2/5 = ?$
- Solve for x: $6x - 2(x + 3) = x - 10$
- How much carpet is required to cover a floor that is 12.5 ft. by 16.2 ft.?

The teachers seemed to conflate level of difficulty with level of understanding. The absurdity of this conflation becomes obvious when we claim that finding the product of two nine-digit numbers assesses a deep and thorough understanding of multiplication. Beginning teachers had particular difficulty in generating items that were other than computational in nature. Further, when eighteen of the teachers were interviewed, they described the teaching of mathematics as a "step by step" process. This, too, suggests a reductionist view of mathematics from which it is difficult to realize the outcomes suggested in the NCTM's *Curriculum and Evaluation Standards.*

We must realize that it is a very tall order for teachers to adopt a relativistic orientation toward knowledge (Perry 1970) in the absence of explicit attention in teacher education programs. We must also realize that the knowledge of mathematics, the teaching and learning of mathematics, and one's orientation toward these types of knowledge are different entities. For example, Wilson (1992) found that the preservice secondary school teachers he studied developed reasonably sophisticated notions about mathematics and mathematical functions but were much less sophisticated about the teaching of functions and mathematics. This finding suggests that teachers need to learn mathematics in a manner that is consistent with the way we expect them to teach. We take too great a leap of faith in assuming that a study of advanced mathematics, of children's misconceptions, of philosophies of mathematics, and of strategies for teaching mathematics can provide, by itself, a foundation from which changes in the teaching of mathematics can be realized. If we want teachers to develop the kind of mathematical communities emphasized in the *Curriculum and Evaluation Standards,* they must participate themselves in mathematical communities in which process is paramount and relativism is revered.

Consider what it means for a teacher to have an understanding of mathematics. Although we are interested in teachers' ability to solve such equations as $4 - 2x = x + 5$, $2^x = 135$, and even $x = 100 \sin x$, the real issue is the means by which they conceptualize the solution process—including the generation of multiple solution methods. For example, solutions to the linear equation can be solved by considering two functions, $f(x) = 4 - 2x$ and $f(x) = x + 5$, and noting, by using a spreadsheet or a graphing calculator, where the function values are the same. Although such an approach may be less efficient, it represents a flexibility that is desired in the teaching of mathematics. This flexibility is essential when solving the third equation, $x = 100 \sin x,$ or other equations for which a standard solution procedure does not exist.

In recent interviews with preservice teachers who had just begun their formal study of mathematics education, we found that many of them lack this flexibility. In general, they have not conceptualized various ways of solving equations. Quite typically, if the students could not remember how to solve an equation (e.g., $2^x = 135$), they were unable to invent solution methods. Although not surprising in itself, this behavior signals a concern that flexible thinking is not likely to occur without being explicitly addressed.

Another component of a teacher's knowledge is the ability to understand students' thinking when responding to various questions. I mentioned earlier that approximately one-half the experienced teachers surveyed by Cooney (1992) were unable to generate items that went beyond computational or simple one-step problems. We also have found that many teachers have difficulty conceptualizing different levels of students' responses other than the degrees to which the students can follow a known procedure. The teachers' analyses tended to be based on a procedural orientation to the task. Further, our experience in having secondary school preservice teachers interview other education students about their understanding of functions suggests that the teachers lack a conceptual basis for analyzing what the students are telling them about functions. This finding suggests another important area for teacher education, namely, to help teachers develop a conceptual basis for understanding how students learn and think about mathematics, as well as opportunities to analyze students' thinking in several contexts.

If teachers only needed knowledge about mathematics, about various ways of interpreting topics in school mathematics, and about how students interpret or construct mathematics, the task of teacher education would be greatly simplified: Give the teachers more of the knowledge they need. The difficulty is with the notion of giving. When the preservice teachers we have studied indicate that a teacher is like a news broadcaster or when they write in their journal entries (in response to a question about listening to students), "I find it difficult to listen to students because I am very worried about what I am going to say next to the class," we begin to sense the magnitude of the problem. Perhaps teacher education conceived as the giving of "more" would provide a better basis from which teachers could then give to their students what they have themselves received. From this perspective, adaptation is not an issue; indeed, adaptation is considered the student's responsibility and not the teacher's. This notion of teacher education is inconsistent with von Glasersfeld's (1989) notion of adaptation.

The ability to become adaptive is the elusive component in many teacher education programs. What kinds of knowledge are necessary to become an adaptive teacher? Certainly one needs to acquire a significant amount of knowledge about mathematics and of the teaching and learning of mathematics, for adaptation cannot be built on pillars of sand that are subject to collapse when faced with either the winds of change or stormy days in the classroom. The ability to be adaptive is rooted in one's orientation toward knowledge. Schon's (1983) reflective practitioner is more concerned about reflection, adaptation, and process than about knowledge per se. This orientation has as much to do with *how* teachers learn to teach mathematics as it does with *what* they learn about teaching mathematics.

THE THIRD DIMENSION OF TEACHER EDUCATION

Treffers (1987) discusses three dimensions, or types, of goals for mathematics education. The first dimension involves very general goals (e.g., critical thinking); the second involves goals that are more content specific. Treffers maintains that an appreciation of these first two dimensions can be realized only when they are set in a didactical context—his third dimension of goals. Treffers's third dimension displays a certain similarity to the idea of context conveyed in this article. Earlier, I discussed the importance of teachers' knowledge of mathematics and the teaching and learning of mathematics. But these types of knowledge become more powerful when cast in a relativistic orientation (Perry 1970) in which context shapes the nature and use of that knowledge. Change the context and we change the nature of the knowledge. Bauersfeld (1988) talks about fundamental relativism as a way of understanding our world; that is, knowledge and understanding are necessarily contextual. The notion of context should not be lost as we envision the kind of knowledge we think teachers should have if they are to become adaptive agents in the classroom.

Recent calls for reform have emphasized the notion of students' mathematical power—the ability to draw on whatever knowledge is needed to solve problems. It might be interesting to consider the similarity between mathematical power for students and pedagogical power for teachers. If mathematical problem solving has to do with recognizing the conditions and constraints of a problem, then pedagogical problem solving has to do with recognizing the conditions and constraints of the pedagogical problem being faced. The context determines how the problem is addressed, not to mention the type of intellectual resources needed to address the problem. To illustrate the importance of pedagogical power, or lack thereof, consider the case of Fred (Cooney 1985), who believed that the essence of mathematics was problem solving yet held a dualistic and limited orientation toward the teaching of mathematics. When his students failed to buy into his orientation toward mathematics and the teaching of mathematics, he had few alternatives to draw on and consequently retreated to teaching by the textbook. The third dimension of teacher education involves helping teachers like Fred develop a relativistic orientation toward knowledge (Perry 1970) so that it can be powerfully adapted to whatever context they find themselves in.

How, then, can we promote the acquisition of this third dimension of teacher education? Over a decade ago, Bauersfeld (1980, p. 38) suggested that teaching is a social activity and that consequently learning to teach is also essentially a social activity:

> But, if we form our cognition and behavior about teaching through social situations, then we can also change this formed cognition and behavior through social situations. We learn to behave in social settings only through the [reflective] participation and action in social settings. Similarly, a teacher will learn to teach or to change his teaching pattern only through [reflective] teaching. Yet, this is not the ruling model of present pre-service teacher training.

The question then becomes, How can we create contexts in teacher education programs that promote the kind of reflection and adaptive behavior that allow teachers to become pedagogically powerful?

PROMOTING PEDAGOGICAL POWER

A reasonable question that we can ask ourselves is, What kind of experiences do teachers need to have in order to develop both the conceptual base and the orientation to become pedagogically powerful? It would seem that teacher education programs ought to have features that—

- enable teachers to develop a knowledge of mathematics that permits the teaching of mathematics from a constructivist perspective;
- offer occasions for teachers to reflect on their own experiences as learners of mathematics;
- provide contexts in which teachers develop expertise in identifying and analyzing the constraints they face in teaching and how they can deal with those constraints;
- furnish contexts in which teachers gain experience in assessing a student's understanding of mathematics;
- afford opportunities for teachers to translate their knowledge of mathematics into viable teaching strategies.

On the surface, there is not much new here. The newness, however, comes in the way in which these features are conceived. To begin, we can engage teachers in the process of connecting with their own learning experiences. The reflection process begins with an analysis of personal classroom anecdotes and with the implications of that analysis for their students' experiences. No magical way exists to promote reflection, but various possibilities include the teachers' keeping a journal about their present learning experiences; writing a short biography about their learning experiences; or writing an essay about how they see mathematics depicted in print, movies, television, and cartoons or how mathematics is used (correctly or incorrectly) in everyday language. Another technique is to propose five or six approaches to the teaching of specific topics (e.g., functions, Pythagorean theorem) and have the teachers allocate a total of 100 points to the approaches according to their judgment of each technique's merit. This assignment of points can then serve as a basis for discussions on what is considered really valuable in teaching that topic. The approaches might vary from an emphasis on students' developing basic skills to an emphasis on connecting mathematics to the real world.

It is particularly important for preservice teachers to see themselves as emerging professionals rather than as accumulators of knowledge, which so typically describes much of their undergraduate experience. One way to break the ice is to engage them in a pedagogical problem in which a teacher has to deal with students' learning difficulties or justify to parents why a particular curriculum is being adopted. Teachers at all levels can be involved in analyzing classroom transcripts or tapes in an effort to understand why certain classroom events occur as they do. Many excellent suggestions are given throughout this book for engaging teachers in problems associated with the teaching of mathematics. (Some of the activities described in this section were developed as part of the National Science Foundation teacher preparation project Integrating Mathematics Pedagogy and Content in Preservice Teacher Education [TPE-9050016].)

Teachers' mathematical experiences must also have a reflective and adaptive orientation. A sample activity that we have used with preservice teachers is given in figure 2.1. Its purpose is to help teachers conceptualize the behaviors of different types of mathematical functions in a manner that is consistent with an exploratory mind-set. The activity is based on Kelly's (1955) imp grid technique used in research in psychology. The primary purpose is to give teachers a context in which they have not only to understand the mathematics by justifying their particular classification system but also to become familiar with a technique that can be adapted to other topics as well, for instance, linear equations or proving triangles congruent. It is essential that teachers see the activity as a means of classifying mathematical objects—a basic component of mathematical thinking.

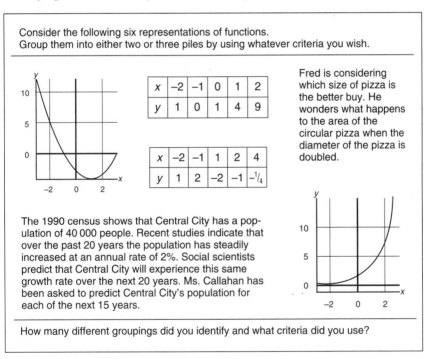

Fig. 2.1. A sample activity

The importance of having teachers develop various ways of solving equations was mentioned earlier. Clearly technology plays a major role in this conceptualization process as we encourage teachers to use technology to solve equations, particularly where standard procedures do not exist.

Another aspect of understanding mathematics is to envision the relationship between mathematics and real-world phenomena. Historically, the competing idealisms have been advanced of mathematics as a science unto itself and as a science that describes the world in which we live. Indeed, it is worthwhile for teachers to consider this debate because it plays a role in the teaching of mathematics. Witness the shift from the emphasis on logical structure that dominated the 1960s to the current emphasis on applications. Many experiments can yield mathematical relationships. But what must be considered is why these experiments should be done in the first place. What purpose do they serve? What is there about the process of collecting data and then "mathematizing the data" that makes it a worthy mathematical activity for students? Without addressing this question, teachers will likely see experiments or exploratory activities as enjoyable pursuits that are relegated to rainy Friday afternoons, given the usual crush of covering curricula. It is not difficult to guess what happens to activities that teachers see as ancillary to the "real curriculum."

In a previous section, I pointed out that teachers, especially preservice teachers, have difficulty in identifying different levels of student responses to a given problem or task. Part of the problem stems from their lack of experience in seeing how students conceptualize mathematics. Teachers should have experience in conducting clinical interviews with students about their mathematical thinking. Sometimes this process can be facilitated by using a nontraditional item such as that in figure 2.2, which has no standard method of solution. As a consequence, the temptation is minimized to describe students' responses as the degree to which a known algorithm is applied correctly. The item requires that the teachers deal with the item conceptually before they can hypothesize what students' responses might be.

A researcher asked many students two questions: "What was your grade on your last math exam?" and "How many hours each night did you usually spend on math homework?" The researcher then sorted students into groups according to how much time they spent on homework. Finally, the researcher computed an average math grade for each of these groups and plotted the averages in the graph on the right.

Write a plausible explanation for the data.

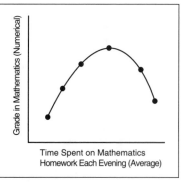

Fig. 2.2. (From the State of Massachusetts Mathematics Assessment Program, Elizabeth Badger, director)

Finally, how does one encourage teachers to listen to what their students are telling them? Without good listening skills, making any meaningful adaptations in the classroom is very difficult. One of the techniques that we have used is to identify a controversial topic, for example, letting students at all grade levels use calculators, and to have pairs of students debate the issue. The catch is that if Jack and Jill are debating, each person can make only one point at a time. If Jack goes first, he can make only one point. But before Jill can make her point, she must paraphrase what Jack said *and* Jack must agree that Jill has accurately characterized his position. They then repeat the process: Jill makes her point, but Jack must have Jill's approval of his characterization of her point before he can proceed. Although artificial in some sense, this exercise forces a person to listen to what the other person is saying; otherwise, the second person is prohibited from making his or her argument. The activity minimizes the I-can't-listen-because-I-am-rehearsing-what-I-am-saying-next effect.

CONCLUSION

Teacher education is not a matter of educators' informing prospective teachers of the Holy Grail for teaching mathematics. We can't meaningfully tell students that *Gone with the Wind* is a classic. They must see that for themselves. They must sense the human drama, the unfolding new order in the South. Similarly, teachers must develop their own sense of what constitutes change in their classrooms. It is counterproductive for teacher educators to try to give teachers the vision that we deem appropriate. Our task, rather, is to help teachers develop professionally so that they can envision that the bridge to a different classroom is close and crossable. When the bridge to reform is deemed unsafe or unreachable, we have laid the groundwork for turning another well-meaning reform into a sea of disappointment.

In a sense, we are asking teachers to become scientific. By this I mean that they learn to identify the constraints they face every day and to hypothesize and test conjectures for dealing with those constraints. It follows that teacher education programs should provide contexts to encourage this "scientific approach" to teaching that honors observing students; hypothesizing about, and examining the effects of, various teaching strategies; and reformulating hypotheses about students' learning.

Fundamentally, teacher education should be grounded in an epistemology that emphasizes solution methods, not solutions. Or to say it in another way, the solution is a process—the search for making a better classroom. Such an orientation suggests that progress is incremental and involves a continuous cycle of teaching and reflecting and that the existence of any single road leading to utopia is only a mirage.

Visions play an integral role in our search for adaptive methods of teaching mathematics and for reflective teacher education programs. They serve as a compass for making practical decisions, as suggested by Barclay (1975). It is through the exercise of adaptation that visions can be realized. Central to this process are the basic tenets of relativity and reflection, rooted in the principle of adaptation and based on the kinds of knowledge previously mentioned. Our goal as teacher educators is to promote adaptation as commonplace in the classroom rather than as an aberration. Indeed, this is the challenge we face in dealing with the practical art of creating innovative mathematics teacher education programs.

REFERENCES

Ball, Deborah L. "Prospective Teachers' Understanding of Mathematics: What Do They Bring with Them to Teacher Education?" Paper presented at the annual meeting of the American Educational Research Association, New Orleans, April 1988.

Barclay, William. *The Gospel of Luke.* Rev. ed. Daily Study Bible: New Testament. Philadelphia: The Westminster Press, 1975.

Bauersfeld, Heinrich. "Hidden Dimensions in the So-Called Reality of a Mathematics Classroom." *Educational Studies in Mathematics* 11 (February 1980): 23–41.

————. "Interaction, Construction, and Knowledge: Alternative Perspectives for Mathematics Education." In *Perspectives on Research on Effective Mathematics Teaching,* edited by Douglas A. Grouws, Thomas J. Cooney, and Douglas Jones, pp. 27–46. Reston, Va.: National Council of Teachers of Mathematics, and Hillsdale, N.J.: Lawrence Erlbaum Associates, 1988.

Begle, Edward G. "Curriculum Research in Mathematics." In *Research and Development toward the Improvement of Education,* edited by Herbert J. Klausmeier and G. T. O'Hearn, pp. 44–48. Madison, Wis.: Dembar Educational Research Services, 1968.

Brown, Stephen I., Thomas J. Cooney, and Douglas Jones. "Mathematics Teacher Education." In *Handbook of Research on Teacher Education,* edited by W. Robert Houston, Martin Haberman, and John Sikula, pp. 639–56. New York: Macmillan Publishing Co., 1990.

Bush, William. "Preservice Secondary Mathematics Teachers' Knowledge about Teaching Mathematics and Decision-Making during Teacher Training." (Doctoral dissertation, University of Georgia, 1982.) *Dissertation Abstracts International* 43 (1983): 2264A.

Cooney, Thomas J. "A Beginning Teacher's View of Problem Solving." *Journal for Research in Mathematics Education* 16 (November 1985): 324–36.

————. *A Survey of Secondary Teachers' Evaluation Practices in Georgia.* Athens, Ga.: University of Georgia Press, 1992.

Eisenberg, Theodore A. "Begle Revisited: Teacher Knowledge and Student Achievement in Algebra." *Journal for Research in Mathematics Education* 8 (May 1977): 216–22.

Fisher, Linda C. "Strategies Used by Secondary Mathematics Teachers to Solve Proportion Problems." *Journal for Research in Mathematics Education* 19 (March 1988): 157–68.

Fitzsimmons, Stephen. *Models of In-Service Education in Mathematics.* Cambridge, Mass.: Abt Associates, 1991.

Graeber, Anna, Dina Tirosh, and Roseanne Glover. "Preservice Teachers' Beliefs and Performance on Measurement and Partitive Division Problems." In *Proceedings of the Eighth Annual Meeting of the North American Chapter of the International Group for the Psychology of Mathematics Education,* edited by Glenda Lappan and R. Even, pp. 262–67. East Lansing, Mich.: Michigan State University, 1986.

Helms, James M. "Preservice Secondary Mathematics Teachers' Beliefs about Mathematics and the Teaching of Mathematics: Two Case Studies." Doctoral dissertation, University of Georgia, 1989.

Kelly, George A. *The Psychology of Personal Constructs.* Vol. 1, *A Theory of Personality.* New York: W. W. Norton & Co., 1955.

Kesler, Reuben. "Teachers' Instructional Behavior Related to Their Conceptions of Teaching and Mathematics and Their Level of Dogmatism: Four Case Studies." Doctoral dissertation, University of Georgia, 1985.

Kuhn, Thomas S. *The Structure of Scientific Revolutions.* 2nd. ed. Chicago: University of Chicago Press, 1970.

Lappan, Glenda, William Fitzgerald, Elizabeth Phillips, Mary Jean Winter, Perry Lanier, Ann Madsen-Nason, R. Even, B. Lee, J. Smith, and D. Weinberg. *The Middle Grades Mathematics Project: The Challenge: Good Mathematics—Taught Well.* Final report to the National Science Foundation for grant no. MDR8318218. East Lansing, Mich.: Michigan State University, 1988.

McGalliard, William A. "Selected Factors in the Conceptual Systems of Geometry Teachers: Four Case Studies." (Doctoral dissertation, University of Georgia, 1982.) *Dissertation Abstracts International* 44 (1983): 1364A.

Mayberry, Joanne. "The van Hiele Levels of Geometric Thought in Undergraduate Preservice Teachers." *Journal for Research in Mathematics Education* 14 (January 1983): 58–69.

Meyerson, Lawrence. "Conceptions of Knowledge in Mathematics: Interactions with and Applications to a Teaching Methods Course." (Doctoral dissertation, State University of New York, Buffalo, 1977.) *Dissertation Abstracts International* 38 (1978): 733A.

National Council of Teachers of Mathematics. *Curriculum and Evaluation Standards for School Mathematics.* Reston, Va.: The Council, 1989.

———. *Professional Standards for Teaching Mathematics.* Reston, Va.: The Council, 1991.

Owens, John. "A Study of Four Preservice Secondary Mathematics Teachers' Constructs of Mathematics and Mathematics Teaching." Doctoral dissertation, University of Georgia, 1987.

Perry, William G. *Forms of Intellectual and Ethical Development in the College Years: A Scheme.* New York: Holt, Rinehart & Winston, 1970.

Schon, Donald A. *The Reflective Practitioner: How Professionals Think in Action.* New York: Basic Books, 1983.

Steen, Lynn. "The Science of Patterns." *Science,* 29 April 1988, pp. 611–16.

Treffers, Adrian. *Three Dimensions.* Boston: D. Reidel Publishing Co., 1987.

von Glasersfeld, Ernst. "Constructivism in Education." In *International Encyclopedia of Education,* edited by Torsten Husen and N. Postlethwaite, pp. 162–63. Supplementary vol. Oxford, England: Pergamon Press, 1989.

Wheeler, Margariete Montague, and Issa Feghali. "Much Ado about Nothing: Preservice Elementary School Teachers' Concept of Zero." *Journal for Research in Mathematics Education* 14 (May 1983): 147–55.

Wilson, Melvin R. "A Study of Three Preservice Secondary Mathematics Teachers' Knowledge and Beliefs about Mathematical Functions." Doctoral dissertation, University of Georgia, 1992.

Wittmann, Erich. "One Source of the Broadcast Metaphor: Mathematical Formalism." In *The Dialogue between Theory and Practice in Mathematics Education: Overcoming the Broadcast Metaphor,* edited by Falk Seeger and Heinz Steinbring, pp. 111–19. Bielefeld, Germany: Institut für Didaktik der Mathematik der Universität Bielefeld, 1992.

3

A Framework for the Professional Development of K–12 Mathematics Teachers

Graham A. Jones
Cheryl A. Lubinski
Jane O. Swafford
Carol A. Thornton

IN WHAT has become a decade of reform stimulated by the publication of the *Curriculum and Evaluation Standards for School Mathematics* (NCTM 1989) and the *Professional Standards for Teaching Mathematics* (NCTM 1991), the role of the teacher as the critical variable in improving mathematics instruction takes on even greater prominence. If the reforms being espoused are to be implemented in a pervasive manner in all mathematics classrooms, it is vital that models be developed for staff development that will produce worthwhile and enduring change.

Two important issues are linked to the creation of appropriate models for staff development programs:

- How do we identify potential leaders for promoting change?
- How do we nurture their leadership?

Of particular interest in the matter of identifying potential leaders is the fact that in the United States, over half of all teachers are over forty years of age (National Center for Education Statistics 1991). Otherwise stated, more than one million teachers in the United States have, on the average, approximately fourteen years' teaching experience in diverse educational settings.

Many of these experienced teachers have a strong commitment to, and interest in, mathematics education; have demonstrated effective interpersonal relationships

This article was based on Project TEAMS (grant no. R168A90270), funded by the U.S. Department of Education, Dwight D. Eisenhower National Math and Science Program; Project LINCS (grant no. TPE9150252), funded by the National Science Foundation, Teacher Preparation and Enhancement Program; and Project CICIM, pending with the National Science Foundation. The views expressed do not necessarily represent those of the funding agencies.

with their colleagues and administrators; are well organized yet flexible; and often have been involved in special mathematical activities or on committees at the local or district level. A substantial number of these teachers have taken the initiative to further their own mathematics education through attendance at professional meetings, enrollment in university courses, reading of mathematics education journals, and participation in staff development programs. These facts imply a cadre of well-qualified and experienced teachers who would be in an optimal position to assume key leadership roles in the mathematics reform movement.

The second issue—nurturing leadership—is more complex. It relates to the design and implementation of effective professional development programs for potential leaders in mathematics education. If these programs are to achieve the desired effects, they must create in potential leaders that "margin of excellence" that will enable them to engender in their colleagues the spirit of reform encapsulated in the *Curriculum and Evaluation Standards* (NCTM 1989) and the *Professional Teaching Standards* (NCTM 1991).

There has been no shortage of professional development programs, nor have these programs been lacking in innovation. The key, however, to creating effective staff development programs is to identify and integrate those program elements that have proved most successful.

Further, in line with recent research on adult learning, it is crucial that programs be designed in ways that take cognizance of the stages of concern experienced by adults in coping with, adapting to, and promoting change. Accommodating the needs of adult learners cannot be ignored in any professional development program that expects to effect long-term change.

Accordingly, this article addresses program design and implementation for staff development in mathematics. In particular, it generates and illustrates an integrated model of professional development that also takes cognizance of the stages of concern typically experienced by teachers facing change.

STAGES OF CONCERN

Hord et al. (1987) note that research has identified seven stages of concern for the typical adult learner. These stages are closely linked to personal beliefs about instructional practice and student learning:

1. *Awareness:* "I am not concerned about any change."

At this stage, teachers are not really concerned about making changes in their approach to instruction, nor are they involved in any innovative programs that might lead to change.

2. *Informational:* "I would like to know more."

Teachers have a general awareness of the innovation and are interested in learning more details about it. They have no personal commitment or anxiety about the innovation. It is largely an interested but detached feeling.

3. *Personal:* "How will using these ideas affect me?"

Teachers are more committed to the innovation and more interested in being involved. They are uncertain, however, about the demands of the innovation and their own ability to meet those demands. This uncertainty can relate to benefits, locus of control, potential conflicts, and even financial implications.

4. *Management:* "I am overwhelmed. How do I organize?"

Teachers are firmly committed to the innovation, but their attention is concentrated on the processes and tasks involved in implementing change and on how best to use both the information and the resources that have been provided. Uppermost in their minds are issues related to efficiency, organization, scheduling, and time management.

5. *Consequence:* "How is the innovation affecting my students?"

Teachers begin to focus on the impact that the change is having on their students. Of particular concern are the relevance of the innovation for students' learning and assessment and how to increase students' performance.

6. *Collaboration:* "I'm concerned about sharing ideas for change."

Teachers have a need to collaborate on the coordination of the program for change and are eager to share ideas about how best to generate change through the innovation.

7. *Refocusing:* "I'm confident I that can improve on ideas learned."

At this ultimate stage, teachers explore more universal benefits from the change, including the possibility of modifying the approach with a more powerful alternative. On the basis of their experience, they have definite ideas for creating such alternatives.

Teachers move from more personal concerns to concerns associated with management, with students, with other teachers, and finally with ways of putting their personal imprimatur on the innovation. Indeed, this latter stage should be the goal of professional development programs that strive to nurture teacher-leaders.

MODELS OF STAFF DEVELOPMENT

According to the National Staff Development Council (Sparks and Loucks-Horsley 1990), five different models of effective staff development for teachers can be identified. Each is based on different assumptions and theoretical and research underpinnings.

Training

Teachers are good learners. The training model of staff development, which may be synonymous with staff development in the minds of many, involves teachers' acquiring knowledge or skills through instruction. Training might include these elements:

- Explanation of a theory
- Demonstration or modeling of a skill
- Practice of the skill
- Feedback about performance

The training model is based on the assumptions that enhanced knowledge influences teachers' beliefs, that some teaching practices are worthy of replication, and that teachers can modify their instructional strategies. Training is usually designed for teachers in one of the first three stages of concern: awareness, informational, and personal.

In training activities, someone, usually an administrator, selects the content, objectives, schedule, and presenters of the training sessions. However, most educators now advocate involving participants in the planning process, including identifying the content, objectives, and presenters and scheduling the format.

Knowledge-level objectives can be effectively achieved through a training program (Wade 1985). A training program is most effective, however, when the program involves theory, demonstration, practice, and feedback (Showers, Joyce, and Bennett 1987). For skill-level objectives, nonthreatening peer-observation activities boost the effectiveness of the training program (Sparks 1986). Training that consists of multiple spaced sessions, allowing teachers the opportunity to practice or be coached in the interval, are more effective than one-shot programs (Duffey and Roehler 1986).

Training programs have been used effectively in mathematics education to change teachers' classroom practice and to improve students' mathematics performance. Examples include training in the effective use of direct instruction (Good and Grouws 1987), cognitively guided instruction (Fennema, Carpenter, and Peterson 1989), and cooperative learning (Davidson 1990). Sparks and Loucks-Horsley (1990) concluded that "under the appropriate conditions, training has the potential for significantly changing teachers' beliefs, knowledge, behavior, and the performance of their students" (p. 15).

Individually Guided Staff Development

Teachers can and do learn many things on their own, and this learning can have a subtle effect on their knowledge. Individually guided staff development is based on the assumption that adults learn most efficiently and will be best motivated when they select their own learning goals and tasks. Since teachers may be at various stages of concern, individually guided staff development can accommodate people with different professional needs.

Individually guided staff development consists of several phases:

- Identification of a need
- Development of a plan
- Implementation of the planned learning activity
- Assessment of the outcome

These phases can be undertaken formally or informally and can consist of a single session (a workshop) or occur over time (reading on the effectiveness of small-group instruction). Although research on the impact of an individualized approach to staff development is largely anecdotal, one study reported that it produced the greatest effect of all the formats examined (Wade 1985).

The most well known examples of individually guided staff development are the activities sponsored by many teachers' centers. Other examples are grants to pursue a special project, such as those funded by Mary Dolciani Scholarship Awards (Mathematics Education Trust) or Eisenhower grants (U.S. Department of Education).

Observation and Assessment

Constructive feedback is a powerful catalyst for change. However, most teachers receive little or no feedback on their classroom performance. The observation-and-assessment model of staff development refers to the process of collecting and assessing objective data on instructional performance through observation. It can take many forms, including peer observation, coaching, clinical supervision, and performance evaluation. Successful observation-and-assessment models include the following:

- A preobservation conference
- A focused observation
- An analysis of the data
- A postobservation conference
- An evaluation of the observation and assessment process (when appropriate).

The observation-and-assessment staff development model is based on the assumption that reflection and analysis are central to professional development and that reflection can be enhanced by another's feedback. Further, teachers are more inclined to gain new knowledge that influences instructional decision making when they receive feedback on their efforts (Lampert 1988). For teachers participating in observation-and-assessment staff development, the stages of concern most likely to dominate are personal, management, and consequence.

The most often used observation-and-assessment development program is clinical supervision of preservice teachers. For practicing teachers, observation and assessment often occur only in conjunction with the evaluation process, in which little attention is given to development. Peer observation and coaching are the most powerful forms of observation-and-assessment staff development and the least used, since few teachers are allocated the time to observe or coach a colleague.

The research on the observation-and-assessment model suggests that teachers' beliefs and behaviors can be changed by the use of this model of staff development. However, evidence is growing that coaching may not be as effective as initially thought and that nonthreatening observation may produce superior results (Sparks 1986; Wade 1985). Although the research suggests that observation-and-assessment staff development is effective in changing the behavior of teachers, no studies have been conducted on the effects on students' performance.

Involvement in a Development or Improvement Process

This model focuses on the combination of learnings that result from the involvement of teachers in curriculum-development or school-improvement projects. This model incorporates—

- the identification of a problem or need;
- the development of an action plan;
- the acquisition of the knowledge or skills required to implement the plan;
- the implementation of the plan or the development of the curricular product;
- the assessment of the outcome.

The involvement-in-change model is based on the assumptions that adults learn most effectively when they have a need to know or a problem to solve and that the people working closest to the problem understand what is required. Teachers involved in this kind of staff development are most likely at the collaboration stage of concern but may pass through other stages during the course of the project.

Curriculum-development or school-improvement projects take many forms, from the adoption of a new textbook to the development of a plan to implement the NCTM's (1989) *Curriculum and Evaluation Standards*. A by-product of involvement in these projects is teachers' own professional development. Research indicates that when support is adequate for curriculum-development or school-improvement projects, the companion outcomes of staff development are achieved (Sparks and Loucks-Horsley 1990).

Inquiry

The inquiry model of staff development requires that teachers identify an area of instructional interest, collect data, and make changes in their instruction on the basis of knowledge gained from interpreting the data. The inquiry model can take many forms from a solitary activity to one carried out by a small group or an entire school. Inquiry may occur in a classroom or at a teachers' center or result from a university class. It can involve reviewing existing data or collecting original data. All inquiries should involve these steps:

- Identification of a problem
- Collection of data
- Analysis and interpretation of the data
- Implementation of changes based on the interpretations
- Evaluation of the effects of the intervention

The inquiry model is based on the assumption that teachers are intelligent, inquiring individuals who can formulate and answer valid questions about their classroom by collecting data and reflecting on them. Teachers enter the inquiry model at the informational stage of concern, wanting to know more about some topic. During the process, they can pass through a number of other stages of concern, perhaps even concluding at the refocusing stage, where they create their own alternative. A number of research studies have demonstrated changes in teachers when they become involved in research (Rauth 1986). Most notable in mathematics education is the Cognitively Guided Instruction Program developed at the University of Wisconsin—Madison (Fennema, Carpenter, and Peterson 1989).

In summary, five models for staff development can be identified. It is, however, possible to integrate aspects of several of the models within one staff development program. Indeed, the significance of these five models is that they can furnish a framework for identifying the key components of an effective staff development program.

AN INTEGRATED FRAMEWORK FOR STAFF DEVELOPMENT

The major goal of staff development programs in mathematics is to effect worthwhile and enduring change in mathematics teaching and learning in elementary, middle, and secondary schools. We anticipate that this change will result from the enhancement of teachers' knowledge, which in turn can be expected to influence their beliefs about, and practices in, mathematics teaching and learning. For change to be widespread, activities should be designed to enable participants to become lead teachers or agents of change at all school levels. Toward this end, we must nurture participants' confidence and competence in teaching mathematics. The ultimate goal is for elementary, middle, or secondary school teachers to be capable of planning and implementing contemporary mathematics programs consonant with the *Curriculum and Evaluation Standards* (NCTM 1989) and the *Professional Teaching Standards* (NCTM 1991) and with current research recommendations.

Professional development projects in mathematics have been initiated in the mathematics department at Illinois State University with these expressed goals. Each project was based on the models identified by Sparks and Loucks-Horsley (1990). We found it very compelling to integrate aspects of the five models in developing the key components of our programs.

Components of an Integrated Program

On the basis of our experience, we present the following six essential components of an integrated staff development program.

> **Involvement in general program design and decision making.** Teachers need to be involved in the planning and execution of staff development programs that are targeted for them.

When our projects were initially formulated, a representative group of administrators and teachers from the participating school districts worked with the project directors to design the program and key components. This involvement in program design was carried into the project activities. For example, in two of the projects, teachers were involved with the staff in the coplanning and copresentation of seminars during the academic year.

> **Instructional activities.** Programs should be multiyear, should include both summer and academic-year components, and should focus on enhancing knowledge of both content and methods.

We have found that three-year programs with academic-year and summer components are effective. In two of the projects, six academic-year seminars were conducted for each of the grade levels. These seminars usually focused on enhancing teachers' pedagogical knowledge. Each summer, workshops or courses led by the staff were offered. These courses usually focused on enhancing teachers' content knowledge but also enabled teachers to explore in depth topics from academic-year seminars.

The NCTM's *Curriculum and Evaluation Standards* (1989) and *Professional Teaching Standards* (1991) were the focus of all the components. Further, in both the summer and academic-year components, a conscious effort was made to enhance teachers' awareness of the implications of research for teaching and learning mathematics. In one project, a separate seminar, Research into Practice, was conducted to embellish the summer content course.

> **School-based activities.** In order to ensure the integration of new knowledge into instructional practice, school-based activities should be part of the staff development program.

In our projects, we have asked teachers to develop instructional units, make and analyze videotapes of their teaching, and keep for specified periods of time journals or logs of their reflections on their experiences. In some instances, assignments were given during the academic-year seminars to be carried out by participants in their own classrooms and then discussed at a subsequent seminar.

As part of the district's contribution, participants have been given monthly released time for collegial collaboration within schools. Involving teams of participants from the same building allowed for peer observation and coaching. Support for the school-based activities was provided by the program staff through on-site visitation.

> **Reflective assessment tasks.** On the basis of the belief that change occurs only through reflection and self-assessment, programs should include structured activities to foster these habits.

In our projects, teachers have used journals to reflect on how information from seminars stimulated changes in their mathematics instruction and in how they assessed learning in their classrooms. The videotaping and analysis of classroom lessons have afforded further opportunities for reflective self-assessment. From viewing videotapes that were kept for the duration of a multiyear program, we found that the change that had occurred was both dramatic and illuminating.

> **Leadership tasks.** Many experienced teachers with the potential to be lead teachers have never been in a leadership role, nor have they envisioned themselves in such a role. To cultivate leaders, a variety of leadership opportunities must be provided.

To furnish leadership opportunities at the school level, we began by having participants interact with other teachers in their building about mathematics teaching. Then they would observe and give feedback to their colleagues and present demonstration lessons in their own classrooms for other teachers to observe. The culminating activity was the development of a school-based curricular plan. In one program, teachers also served as mentors to preservice teachers, teaming with them to plan and present mathematics lessons.

In order to provide leadership tasks that went beyond the school, teachers were involved in collaborative curriculum planning with neighboring schools and were encouraged to make presentations at professional meetings and conferences. The coplanning and copresenting of the academic-year seminars by participants offered further opportunities to develop confidence in leading seminars. In several instances, with staff support, teachers' ideas were expanded into articles that were submitted to the NCTM's journals.

> **Evaluation.** Effective staff development programs need ongoing monitoring and in-depth evaluation.

In all our projects we have solicited both formal and informal feedback on individual components of the programs and have modified the programs accordingly. To determine the impact of a project, we have gathered longitudinal data on the time spent on various content topics, on teaching emphases, and on beliefs about teaching and learning as well as students' performance. In some programs, case studies were carried out with selected participants.

Relationship of the Models to an Integrated Program

The integrated program for staff development presented above employed elements from each of the five models of effective staff development identified by Sparks and Loucks-Horsley (1990). As illustrated in figure 3.1, the five models of staff development are interfaced with the program components as follows.

Training

Academic-year seminars and summer courses and seminars were organized to foster new understandings and formulate new perspectives about mathematics teaching and learning. The vision of training described by Sparks and Loucks-Horsley (1990) emphasized explanation and demonstration. Although our programs involved some of these methods, we adopted a broader perspective on training. Accordingly, our training activities included explaining and discussing theory and new content, reporting research findings, sharing ideas among participants, and modeling and analyzing instructional activities. Teachers were encouraged to try out new ideas in their classrooms and to report on the outcomes. Peer coaching, videotaped lessons, reflective journals, and staff observations supplied feedback that enabled teachers to monitor their own development.

Individually guided staff development

Individual staff development needs were accommodated in a number of ways. For example, in one project, participants developed their own instructional units with the assistance of the project's staff. These units varied by theme and intent but were developed in response to teachers' needs. Teachers also selected the seminar that they would copresent and then sought out materials and developed their own presentations. The expectation was borne out that some teachers would devise innovative instructional programs in areas that were critical to their own needs. Some of these experiences resulted in submissions to professional journals and presentations at professional meetings.

Observation and assessment

This component received major emphasis. It was part of the collegial interaction among teachers in the project during monthly released time or classroom visits. It was also part of the reflective process associated with the creation of journals and the self-analysis of videotaped lessons. A rich form of observation and assessment occurred as staff worked very closely with a small number of the teachers identified as subjects for case studies. In a related, distinctive example of the model, teachers and preservice teachers coplanned and copresented mathematics lessons, following a process of peer collaboration and coaching.

Involvement in the improvement process

An application of this model occurred in the systematic, long-term planning for school-based change required of teams from the same school or district. This planning also included subsequent collaborative activities with nonparticipating teachers.

Inquiry

The inquiry model was an important feature of the projects. For example, ongoing analysis of students' thinking provided a basis for instructional decision making. In one project, the inquiry model was also embodied in the summer Research into Practice seminar, in which project teachers synthesized or collected research data, applied their findings to their own instructional programs, and implemented and evaluated the resulting changes. We expected that this type of inquiry would produce a ripple effect within participants' schools and subsequently result in the implementation of more pervasive changes.

An effort was also made throughout the projects to take cognizance of the stages of concern identified by Hord et al. (1987) discussed earlier. In their interactions with the teachers, the projects' staff members were very aware of these stages of development and of the need to be sensitive to the fact that the teachers spanned diverse stages of the continuum. Activities were planned with enough flexibility to allow for different levels of participation and commitment.

During classroom visits, expressions of concern associated with the various stages, for example, personal concerns, management concerns, and concerns about consequences and collaboration, were openly discussed and addressed. Finally, concerns about refocusing were embedded in the very heart of the projects. The long-term goal was to move teachers to the collaboration or refocusing stages, where they would be able to share ideas for change and become confident enough to embellish the ideas.

Program Component	Model				
	Training	Individually Guided Development	Observation & Assessment	Involvement in a Process	Inquiry
Involvement in program design and decision making					
Cooperative formulation				✗	
Coplanning and copresentation		✗		✗	
Instructional activities					
Seminars during school year	✗				
Summer workshops and courses	✗				
Summer research seminar	✗				✗
School-based activities illuminated by reflective assessment					
Development of an instructional unit		✗			
Self-analysis of videotaped lessons			✗		
Peer analysis of a lesson			✗		
Reflective journals			✗		
Classroom visits and support			✗		
Analysis of students' thinking					✗
Leadership tasks					
School plan for change				✗	
Sharing at NCTM conferences and through publications		✗			
Evaluation					
Evaluation and case studies			✗		

Fig. 3.1. The interface between the five models of staff development and the components of the staff development program

CONCLUDING REMARKS

This article has identified five models (Sparks and Loucks-Horsley 1990) that when integrated, have considerable potential for use in designing and implementing staff development programs. In particular, we contend that these models, together with the stages of concern associated with adult learners (Hord et al. 1987), serve as a framework for an effective staff development program designed to cultivate teacher-leaders. We have illustrated how the five models of staff development and the concerns of the adult learner can be integrated into the general design, major components, and evaluation of staff development programs for teachers of grades K–12. Our success suggests that this approach is a viable and transferable general framework that is capable of producing worthwhile and enduring change in mathematics instruction.

REFERENCES

Davidson, Neil, ed. *Cooperative Learning in Mathematics: A Handbook for Teachers.* Menlo Park, Calif.: Addison-Wesley Publishing Co., 1990.

Duffy, Gerald, and Laura Roehler. "Constraints on Teacher Change." *Journal of Teacher Education* 37 (January/February 1986): 55–58.

Fennema, Elizabeth, Thomas P. Carpenter, and Penelope Peterson. "Teachers' Decision Making and Cognitively Guided Instruction: A New Paradigm for Curriculum Development." In *Facilitating Change in Education,* edited by Ken Clements and Nerida F. Ellerton. Geelong, Victoria, Australia: Deakin University Press, 1989.

Good, Thomas L., and Douglas A. Grouws. "Increasing Teachers' Understanding of Mathematical Ideas through In-Service Training." *Phi Delta Kappan* 68 (June 1987): 778–83.

Hord, Shirley M., William L. Rutherford, L. Huling-Austin, and Gene E. Hall. *Taking Charge of Change.* Alexandria, Va.: Association for Supervision and Curriculum Development, 1987.

Lampert, Maggie. "What Can Research on Teacher Education Tell Us about Improving Quality in Mathematics Education?" *Teaching and Teacher Education* 4 (1988): 157–70.

National Center for Education Statistics. *Digest of Education Statistics.* Washington, D.C.: Government Printing Office, 1991.

National Council of Teachers of Mathematics. *Curriculum and Evaluation Standards for School Mathematics.* Reston, Va.: The Council, 1989.

————. *Professional Standards for Teaching Mathematics.* Reston, Va.: The Council, 1991.

Rauth, Marilyn. "Putting Research to Work." *American Educator* 10 (Winter 1986): 26–31, 45.

Showers, Beverly, Bruce Joyce, and Barrie Bennett. "Synthesis of Research on Staff Development: A Framework for Future Study and a State-of-the-Art Analysis." *Educational Leadership* 45 (November 1987): 77–87.

Sparks, Dennis, and Susan Loucks-Horsley. *Five Models of Staff Development.* Oxford, Ohio: National Staff Development Council, 1990.

Sparks, Georgea M. "The Effectiveness of Alternative Training Activities in Changing Teaching Practices." *American Educational Research Journal* 23 (Summer 1986): 217–25.

Wade, Ruth K. "What Makes a Difference in In-Service Teacher Education? A Metaanalysis of Research." *Educational Leadership* 42 (December 1984/January 1985): 48–54.

4

Ten Key Principles from Research for the Professional Development of Mathematics Teachers

Doug Clarke

THE *Curriculum and Evaluation Standards for School Mathematics* (National Council of Teachers of Mathematics [1989]) and the *Professional Standards for Teaching Mathematics* (NCTM 1991) have promoted a vision for teaching and learning that has growing support from the mathematics education community. There is increasing recognition, however, that without carefully planned professional development programs, the chance of widespread implementation of these exciting reforms is small. Support is essential for practicing teachers who express an interest in teaching in a way described in the *Standards,* with all that such teaching implies in terms of content, pedagogy, and assessment. This paper extracts ten important principles from the professional development research literature that can be used to guide the planning and implementation of staff development programs (see fig. 4.1).

For each of these principles, brief support from the literature and the experiences of the author will be presented. A more detailed basis for each of these principles can be found in Clarke (1991). Schlechty (1983) observed that professional development can serve at least three main purposes: (1) an "establishing" function to promote organizational change, (2) a "maintenance" function, and (3) an "enhancement" function to improve the individual teacher's practice. The latter purpose will underpin the content of this paper.

1. Address issues of concern and interest, largely (but not exclusively) identified by the teachers themselves, and involve a degree of choice for participants.

Professional development programs are more likely to achieve significant change in classroom practice if they are seen by teachers as being responsive to their needs. A number of attempts have been made to ascertain teachers' preferences regarding the content, form, and style of professional development programs. These studies assumed that teachers are in the best position to determine

---

Ten Important Principles of Professional Development

1. Address issues of concern and interest, largely (but not exclusively) identified by the teachers themselves, and involve a degree of choice for participants.

2. Involve groups of teachers rather than individuals from a number of schools, and enlist the support of the school and district administration, students, parents, and the broader school community.

3. Recognize and address the many impediments to teachers' growth at the individual, school, and district level.

4. Using teachers as participants in classroom activities or students in real situations, model desired classroom approaches during in-service sessions to project a clearer vision of the proposed changes.

5. Solicit teachers' conscious commitment to participate actively in the professional development sessions and to undertake required readings and classroom tasks, appropriately adapted for their own classroom.

6. Recognize that changes in teachers' beliefs about teaching and learning are derived largely from classroom practice; as a result, such changes will follow the opportunity to validate, through observing positive student learning, information supplied by professional development programs.

7. Allow time and opportunities for planning, reflection, and feedback in order to report successes and failures to the group, to share "the wisdom of practice," and to discuss problems and solutions regarding individual students and new teaching approaches.

8. Enable participating teachers to gain a substantial degree of ownership by their involvement in decision making and by being regarded as true partners in the change process.

9. Recognize that change is a gradual, difficult, and often painful process, and afford opportunities for ongoing support from peers and critical friends.

10. Encourage participants to set further goals for their professional growth.

Fig. 4.1.

their professional development needs. In research conducted by Zigarmi, Betz, and Jensen (1977), the most common forms of in-service training—after-school or one-day regional workshops—were judged to be the least useful. These results correspond with other research that has established the minimal effect on teacher change of the "one shot" model. "Current trends" workshops were rated as the most useful type of activity included in the survey. This finding was important, since many teachers (who can't be expected to be aware of all new initiatives) are not familiar with successful new approaches or trends. A balance between expressed areas of need and emerging issues seems appropriate.

Research has demonstrated considerable differences among teachers in the knowledge, beliefs, and skills they bring to professional development programs and in their personal characteristics, such as states of growth and flexibility of thinking. Doyle and Ponder (1977), using naturalistic methods involving discussions with experienced teachers and with other school personnel during in-service programs and workshops and interviews with students enrolled in graduate courses in education, identified the "practicality ethic" as the main influence on teachers' adoption of innovations. They claimed that in-service messages that are seen as practical would be incorporated, at least tentatively, into teachers' plans. They characterized the bulk of teachers as "pragmatic skeptics," who were strongly guided by the central role of personal preference, immediate contingencies and consequences, and "the concrete and the procedural rather than the abstract and the general.... Teacher behavior, including reactions to change proposals, is seen as an outgrowth of effort to meet environmental demands imposed by the distinctive ecology of the classroom" (pp. 4–5).

The recognition that there is a progression of concerns from *self* (do I know the content?) to *task* (how do I explain this concept?) to *impact* (are they learning something worthwhile?) among teachers involved in innovations was a major feature of the important investigation done by Hall and Loucks (1978) as part of the work on the Concerns-Based Adoption Model (CBAM) at the Research and Development Center for Teacher Education at the University of Texas. Their research focused on diagnosing group and individual needs during the adoption of innovations. Their underlying assumptions included the notions that change is a process, not an event; that the individual must be the primary focus of interventions; and that change is a highly personal process. Teachers, like any professionals, show wide variation across a range of characteristics, and professional development programs should reflect this variation through both the content of programs and the degree of choice afforded participants.

2. Involve groups of teachers rather than individuals from a number of schools, and enlist the support of the school and district administration, students, parents, and the broader school community.

U.S. teachers work largely in professional isolation. A survey of high school teachers found that 46 percent of teachers spent less than one hour a month meeting with colleagues on curriculum and instruction (Moles 1988). Whereas teachers generally believe that their major assistance in maintaining and improving their skills comes from other teachers, it is ironic that they have little opportunity to work with, observe, or receive feedback from, peers on their teaching. Encouraging teachers to participate in professional development programs with colleagues from their own school can start to build the norms of collegiality and experimentation that are necessary for significant change (Little 1982).

Fullan (1990) argued that sprawling and prominent professional development, although rightly recognized as the central strategy for improvement, "is frequently separated artificially from the institutional and personal contexts in which it operates" (p. 4). In arguing that staff developers must work with schools as organizations as much as they work with individuals or small groups of teachers, Fullan identified teacher collegiality and other aspects of collaborative-work cultures as variables related to the likelihood of successful implementation.

McEvoy (1985) identified everyday routine activities and exchanges that the school principal can undertake in the process of monitoring and managing the school that can informally initiate professional development at an individual and group level. The routines include alerting teachers individually to workshop opportunities; disseminating professional materials to individuals and groups, encouraging feedback on the teachers' opinions of the materials; publicly stating themes or areas of emphasis and using them in informal discussions with individuals; seeking answers from individuals to issues of importance to the school; conveying an impression of supporting experimentation and innovation; publicizing individual teachers' achievements; and facilitating collegial interchange.

Two greatly underestimated groups in change initiatives are students and parents (Fullan and Stiegelbauer 1991). As teachers attempt new ideas and approaches in the mathematics classroom, students and parents often react negatively to teaching that moves beyond the "drill and practice" that in their opinion *is* mathematics. Those involved in the leadership and implementation of professional development programs can furnish an impetus to the change process if they share the philosophy and the practical implication of the initiative with parents and students and enlist their support. Both groups have much to offer, and involving them early in the process is far preferable to "putting out fires" later on.

3. Recognize and address the many impediments to teachers' growth at the individual, school, and district level.

As teachers seek to increase their knowledge and expand the boundaries of their "comfort zone," many barriers arise that serve as impediments. Research evidence clearly shows that if those involved in the planning and delivery of professional development programs fail to address these impediments, the chances for desired outcomes are severely reduced. Although recognizing that many more could be stated (Clarke 1991), I list twelve major impediments under four categories:

Impediments external to the school

1. An inadequate theory of implementation
2. The lack of sustained central office support, funding, and follow-up
3. The failure to understand and take into account site-specific differences among schools

Impediments related to school organization and administration and to the school community

4. The lack of time for individual reading and reflection, the lack of joint planning time with other teachers, and the lack of work together in classrooms—all leading to a feeling of professional isolation
5. Student-assessment and teacher-evaluation methods that are not in harmony with the proposed changes

6. The perceptions of students, parents, the principal, the school board, and the immediate community about essential content, appropriate pedagogy, and assessment.

Impediments related to teachers' beliefs, knowledge, and practice

7. The lack of commitment to, and ownership of, the proposed changes
8. The "practicality ethic" held by many teachers (i.e., if a proposed innovation is not seen as practical in terms of classroom implementation, it will be rejected)
9. Teachers' inadequate knowledge, in particular, mathematical content knowledge and pedagogical content knowledge

Impediments related to the content of the staff development sessions

10. The lack of a link between theory and the realities of the classroom
11. An emphasis on correcting deficits rather than encouraging professional growth
12. The lack of incorporation of knowledge about teaching and learning into the content and style of the staff development program

Professional development programs vary in many aspects, including focus, funding, and length of time, and it is unlikely that all these impediments would apply in a given situation. Nevertheless, they serve as a basis for discussion in the planning process for professional development. Although the list of impediments seems so immense that no successful professional development is possible, successful programs *have* been conducted; the purpose of this paper is to identify the key components of professional development programs that have met with success. Situations in which two individuals in the same school have responded to the same staff development program in completely different ways show that it is possible for two teachers to be faced with very similar impediments and yet respond quite differently. Teachers can be overwhelmed by the nature and quantity of impediments to change, but as Griffin (1991) argues, as they learn new ways to work more productively together and singly, they will regard problems and frustrations as a matter of course "rather than [use] the dilemmas and tensions as reasons or excuses to abandon the enterprise" (p. 255).

4. Using teachers as participants in classroom activities or students in real situations, model desired classroom approaches during in-service sessions to project a clearer vision of the proposed changes.

Many authors have advocated in-service programs that offer both experiential learning ("learning by doing") and informal learning situations in which social interaction can occur. Wood and Thompson (1980) stressed the importance of *modeling,* in which participants practice what they are to learn in simulated and real work settings. The emphasis should be on small groups and learning from one another. "The principles and skills developed through experiential learning are more easily recalled because they are tied to a sequence of personal actions and consequences" (Wood and Thompson 1980, p. 377).

Agreement seems to be growing on the principal aspects of the style of activity for an effective in-service program. It is worth making two points in relation to this. First, in-service programs that involve teachers actively are preferred by teachers (Zigarmi, Betz, and Jensen 1977). Second, despite the research basis in adult learning and effective in-service programs, the active approach is far from the dominant style of in-service session. The most common form of staff development, certainly in Australia and the United States, continues to be the one-shot in-service seminar in which an external expert makes a presentation, with little active involvement and no follow-up.

5. Solicit teachers' conscious commitment to participate actively in the professional development sessions and to undertake required readings and classroom tasks, appropriately adapted for their own classroom.

Two aspects of commitment are important here: the commitment to active participation in the program and the commitment to the philosophy and approaches underlying the program. The Australian Mathematics Curriculum and Teaching Program (MCTP) (Lovitt and Clarke 1988, 1989) stressed the importance of teachers' suspending judgment on, say, the value of exploring social issues in the mathematics classroom while making a commitment to try with their own students some of the activities that they had seen modeled in in-service sessions. This classroom experience and the opportunity for reflection on it then provided a basis for further informed discussion of the merits of the approach.

The level of commitment of project teachers to new approaches was found by the Rand Study (McLaughlin and Marsh 1978) to have a highly positive relationship to all the project outcomes (e.g., change in teachers, change in students' performance, and long-term use of methods and materials). Some would argue that commitment is a characteristic of an individual teacher and not particularly affected by policies or program strategies. This view implies that change projects should target only those teachers who exhibit strong initial interest and motivation. However, Huberman and Miles (1984), in a large qualitative study of professional development projects conducted over three years, concluded that commitment *followed* competence. They found that in many instances, teachers who were initially lukewarm about, or wary of, the innovation demonstrated, after receiving appropriate peer and external assistance, a strong commitment that had not been present at the program's commencement.

The literature suggests that flexibility, general motivation to learn, and risk tak-ing are desirable characteristics—more important than initial commitment. McLaughlin and Marsh suggested that teachers' commitment was influenced by the motivation of district managers (evidenced by a concern for schools and classrooms rather than political motives), by project-planning strategies (collab-orative planning was preferred), and by the scope of the proposed project (the more effort required of teachers and the greater the overall change in teaching style attempted by the project, the higher the proportion of committed teachers).

Not surprisingly, commitment is more important when difficulties are encountered early in the implementation of an innovation. This phenomenon expresses itself in two ways: by carrying teachers over the rough spots as they "keep the faith" and by causing principals and central office staff to offer greater support. "In all cases, when initial commitment was low, it tended to be boosted by the presence of district-level and peer assistance, by the experience of mastery and the resulting change in practice, and by gradually improving classroom or building fit" (McLaughlin 1991, p. 268). Not surprisingly, Huber-man and Miles (1984) also found that a great impact on student learning was more likely when the teachers' commitment was high.

6. Recognize that changes in teachers' beliefs about teaching and learning are derived largely from classroom practice; as a result, such changes will follow the opportunity to validate, through observing positive student learning, information supplied by professional development programs.

Most writers acknowledge that professional development programs are about change—in either teachers' knowledge, teachers' beliefs and attitudes, classroom practice, students' learning, or a combination of these. Most teachers are motivated to take on professional development programs because they wish to become better teachers and they believe that their students will benefit. However, much is still to be learned about the *process* of change in the mathematics classroom.

One common approach to professional development is to attempt to create some sort of change in teachers' knowledge, beliefs, and attitudes and trust that such a change will *then* lead to appropriate changes in classroom practice and improved student learning. This approach involves teachers' making some sort of commit-ment to the particular curriculum or teaching method *prior to* classroom practice. Guskey's alternative model (1986) proposed that teachers' beliefs and attitudes are likely to change significantly only *after* they see evidence of changes in students' learning. This model of teacher change rests on various assumptions. Most teachers define their success in terms of the learning of their students rather than in terms of their own actions or other factors. Teachers expect that professional development will provide practical, concrete ideas for their classroom; when professional devel-opment is undertaken in isolation from a teacher's ongoing classroom role, it is likely to have little effect on either teaching practice or students' learning. Beliefs and attitudes about teaching and learning are derived largely from classroom experience, either in the role of teacher or previously in the role of learner (Goldsmith and Nelson 1991). "By interacting with their environment, with all its demands and problems, teachers appear to evaluate and reorganize their beliefs through reflective acts, some more than others" (Thompson 1992, p. 139).

7. Allow time and opportunities for planning, reflection, and feedback in order to report successes and failures to the group, to share "the wisdom of practice," and to discuss problems and solutions regarding individual students and new teaching approaches.

In comparison with teachers in other countries, U.S. teachers have little time to engage in planning, reflection, and feedback, particularly with colleagues (Stigler and Stevenson 1991). This deficiency is a major impediment to their fulfilling their professional role. Thompson (1992) observed that the extent to which experienced teachers' beliefs about the learning of mathematics are consistent with their classroom practice depended largely on their tendency to reflect on their actions, whereby they "gain an awareness of their tacit assumptions, beliefs, and views, and how these relate to their practice.... [They] develop coherent rationales for their views, assumptions, and actions, and become aware of viable alternatives" (p. 139).

Explaining why teachers are often less than precise about their thoughts and considerations, Griffin (1991, pp. 248–49) claimed that their inarticulateness was due to the absence of any need to become articulate, to be communicative, or to use thoughts as objects of systematic attention with their colleagues:

> Why should teachers sometimes be chastised for their lack of eloquence about their implicit theories of instruction, largely unarticulated, when discourse systems that might make those theories explicit are systematically absent from their work settings?

Shulman (1987) presents a positive view of teachers as a scholarly community and salutes "the wisdom of practice—the maxims that guide (or provide reflective rationalization for) the practices of able teachers" (p. 11). However, little opportunity arises for this wisdom of practice to be shared. The major aim of the Australian MCTP project was to capture and share the wisdom of practice through the detailed documentation of exemplary classroom practice and the use of these materials in ongoing professional development (Stephens et al. 1989).

8. Enable participating teachers to gain a substantial degree of ownership by their involvement in decision making and by being regarded as true partners in the change process.

Fenstermacher and Berliner (1985) suggested a framework for classifying staff development programs according to four continua that relate to issues of ownership: (1) how the program was initiated (externally ↔ internally), (2) the purpose of the program (compliance ↔ enrichment), (3) the level of participation (all teachers ↔ one teacher), and (4) the reasons for participation (mandated ↔ voluntary). They concluded that the more bottom-up a profile, the easier it is, in general, to meet the conditions for "valued" professional development. McPherson (1980) stated that teachers ought to be involved at all levels of planning and implementation. He didn't see their involvement as a matter merely of governance but also of growth and learning. "Taking teachers seriously as professionals makes them capable of participating in their own regeneration. It makes it possible, too, for them to choose how they will align their personal and professional goals" (p. 130). In arguing for more interactive staff development, Griffin (1991, p. 248) claimed that "research on teachers' thinking, on teachers' implicit theories, and on expert teachers supports a view of teachers as persons who are thoughtful about their work, their impact on students, their strong points and their vulnerabilities." According to Doyle and Ponder (1977), change projects that bypass decision making by teachers are often unsuccessful in achieving their goals because they mask the operation of the practicality ethic, thereby ignoring the fact that the ultimate fate of the professional development program depends on the decisions of the user—the classroom teacher.

McLaughlin (1991) identified frequent project meetings as important to successful implementation and long-term continuation because they afford teachers an opportunity to influence project decisions and they engender in teachers a sense of ownership. Regular meetings and appropriate assistance from consultants enabled teachers to integrate project practices into their classroom so as to make them "theirs." Offering a range of choice in professional development programs is another important factor in fostering ownership. As discussed earlier, the involvement of students and parents in the change process is likely to lead to a greater sense of ownership by these two important groups.

9. Recognize that change is a gradual, difficult, and often painful process, and afford opportunities for ongoing support from peers and critical friends.

Agreement is growing that significant change is a gradual process over an extended period of time (Fullan 1990), and many writers acknowledge the importance of support and technical assistance at various stages of the change process. Many examples of change projects can be cited that put almost all resources into the in-service aspects of the project but failed to provide appropriate support in the actual implementation. McLaughlin (1991) found that classroom assistance by local resource personnel and outside consultants, project meetings, and teachers' participation in project meetings, when taken together and when seen as useful by school staff, had a major positive effect on project outcomes. Local consultants, because they were available on an as-needed basis and were more likely to provide concrete, situated assistance, were preferred to external consultants.

Day (1985) is one of the few authors to emphasize explicitly the importance of *affective* support in the process of teacher change, although much of the work of others implies its significance. Simon and Schifter (1991) emphasized the need for support for teachers who are struggling with the "emotional load of being a novice" to new approaches and content.

Interestingly, Huberman and Miles (1984) found that a smooth early use by teachers of innovative approaches was a bad sign. An easy implementation seemed to be accomplished by reducing the initial scale of the project and by lowering the gradient of the actual change in practice. "Downsizing got rid of most headaches during the initial implementation but also threw away most of the potential rewards; the project often turned into a modest, if sometimes trivial enterprise" (p. 273). The notion of downsizing is similar to the idea of teachers' reshaping an activity or a program to bring it back into the boundaries of the teachers' "comfort zone," as described by Stephens et al. (1989).

Although downsizing is a cause for concern, it is now increasingly accepted by those responsible for the professional development of teachers that all teachers *adapt* rather than adopt innovative curriculum materials and that they *should* do so (Snyder, Bolin, and Zumwalt 1992). McLaughlin and Marsh (1978, p. 77) described the "process of mutual adaptation in which teachers modify their practices to conform to project requirements and project technologies are adapted to the day-to-day realities of the school classroom." Teachers, like students, cannot be viewed as passive receptors of innovations and must be given the opportunity to make informed decisions about the use of any innovation in their classroom. The ways in which teachers adapt innovations to their own classrooms are influenced by the emphasis in in-service sessions on the underlying principles and meaning behind the suggested changes, the assistance offered by support personnel within the school, and frequent meetings of teachers and project coordinators. Such support in the actual process of use is essential for genuine change to occur.

10. Encourage participants to set further goals for their professional growth.

The recognition that constructive change (Romberg and Price 1983) to a teaching and learning environment advocated by the *Standards* and other reform documents is a slow process implies that teachers should be encouraged to regard their professional development as a continuous process rather than a series of off-and-on experiences. Those responsible for the professional development of mathematics teachers must provide appropriate opportunities for teachers to examine their current practice on a regular basis and to plan appropriately for their future professional growth.

CONCLUSION

This review of the literature leads to a conclusion that an "ideal" professional development program would have incorporated or addressed each of these principles in the planning and implementation of the change process and would therefore have an excellent chance of supporting the professional growth of teachers and the learning of their students.

REFERENCES

Clarke, Doug M. "The Role of Staff Development Programs in Facilitating Professional Growth." Unpublished manuscript, University of Wisconsin—Madison, 1991.

Day, Christopher. "Professional Learning and Researcher Intervention: An Action Research Perspective." *British Educational Research Journal* 11 (1985): 133–51.

Doyle, Walter, and Gerald A. Ponder. "The Practicality Ethic in Teacher Decision-Making." *Interchange* 8 (1977): 1–12.

Fenstermacher, Gary D., and David C. Berliner. "Determining the Value of Staff Development." *Elementary School Journal* 85 (1985): 281–314.

Fullan, Michael G. *The New Meaning of Educational Change.* New York: Teachers College Press, 1991.

———. "Staff Development, Innovation and Institutional Development." In *Changing School Culture through Staff Development,* Yearbook of the Association for Supervision and Curriculum Development, edited by Bruce Joyce, pp. 3–25. Alexandria, Va.: The Association, 1990.

Goldsmith, Lynn T., and Barbara Scott Nelson. "Transforming Teachers: Perception, Practice, and Reality." Paper presented at the Psychology of Mathematics Education Research Presession (Research Methodologies for Studying Teacher Change), October 1991, Blacksburg, Va.

Griffin, Gary A. "Interactive Staff Development: Using What We Know." In *Staff Development for Education in the Nineties: New Demands, New Realities, New Perspectives,* edited by Ann Lieberman and Lynn Miller, pp. 243–58. New York: Teachers College Press, 1991.

Guskey, Thomas R. "Staff Development and the Process of Teacher Change." *Educational Researcher* 15 (May 1986): 5–12.

Hall, Gene E., and Susan Loucks. "Teachers' Concerns as a Basis for Facilitating and Personalizing Staff Development." *Teachers College Record* 80 (September 1978): 36–53.

Huberman, A. Michael, and Matthew B. Miles. *Innovation Up Close: How School Improvement Works.* New York: Plenum Publishing Corp., 1984.

Little, Judith W. "Norms of Collegiality and Experimentation: Workplace Conditions of School Success." *American Educational Research Journal* 19 (Fall 1982): 325–40.

Lovitt, Charles J., and Doug M. Clarke. *Mathematics Curriculum and Teaching Program Activity Bank.* Vols. 1 and 2. Carlton, Victoria, Australia: Curriculum Corp., 1988 (vol. 1), 1989 (vol. 2).

McEvoy, Barbara. "Everyday Acts: Staff Development as Continuous Informal Routine." Paper presented at the annual meeting of the American Educational Research Association, April 1985, Chicago.

McLaughlin, Milbrey W. "Enabling Professional Development: What Have We Learned?" In *Staff Development for Education in the Nineties: New Demands, New Realities, New Perspectives,* edited by Ann Lieberman and Lynn Miller, pp. 61–82. New York: Teachers College Press, 1991.

McLaughlin, Milbrey W., and David D. Marsh. "Staff Development and School Change." *Teachers College Record* 80 (1978): 69–94.

McPherson, Bruce. "Are They in Roses? Teachers and Staff Development." *Journal of Education* 162 (1980): 120–34.

Moles, Oliver C., ed. *High School and Beyond: Administrator and Teacher Survey (1984) Data File User's Manual.* Washington, D.C.: U.S. Department of Education, Office of Educational Research and Improvement, 1988.

National Council of Teachers of Mathematics. *Curriculum and Evaluation Standards for School Mathematics.* Reston, Va.: The Council, 1989.

———. *Professional Standards for Teaching Mathematics.* Reston, Va.: The Council, 1991.

Romberg, Thomas A., and Gary G. Price. "Curriculum Implementation and Staff Development as Cultural Change." In *Staff Development,* Eighty-second Yearbook of the National Society for Study in Education, pt. 2, edited by Gary A. Griffin, pp. 154–84. Chicago: University of Chicago Press, 1983.

Schlechty, Paul, ed. *Understanding and Managing Staff Development in an Urban School System.* Final Report. Washington, D.C.: National Institute of Education, 1983.

Shulman, Lee S. "Knowledge and Learning: Foundations of the New Reform." *Harvard Educational Review* 7 (February 1987): 1–22.

Simon, Martin A., and Deborah Schifter. "Towards a Constructivist Perspective: An Intervention Study of Mathematics Teacher Development." *Educational Studies in Mathematics* 22 (1991): 309–31.

Snyder, Jon, Frances Bolin, and Karin Zumwalt. "Curriculum Implementation." In *Handbook of Research on Curriculum,* edited by Philip W. Jackson, pp. 402–35. New York: Macmillan Publishing Co., 1992.

Stephens, W. Max, Charles J. Lovitt, Doug M. Clarke, and Thomas A. Romberg. "Principles for the Professional Development of Teachers of Mathematics." In *School Mathematics: The Challenge to Change,* edited by N. Ellerton and M. A. Clements, pp. 220–49. Geelong, Victoria, Australia: Deakin University Press, 1989.

Stigler, James W., and Harold W. Stevenson. "How Asian Teachers Polish Each Lesson to Perfection." *American Educator* 15 (Spring 1991): 12–47.

Thompson, Alba. "Teachers' Beliefs and Conceptions: A Synthesis of Research." In *Handbook of Research on Mathematics Teaching and Learning,* edited by Douglas A. Grouws, pp. 127–46. New York: Macmillan Publishing Co., 1992.

Wood, Fred H., and Steven R. Thompson. "Guidelines for Better Staff Development." *Educational Leadership* 37 (February 1980): 374–78.

Zigarmi, Patricia, Loren Betz, and Darrell Jensen. "Teachers' Preferences in and Perceptions of In–Service Education." *Educational Leadership* 34 (April 1977): 545–51.

5

Building Professional Development into the Culture of Schools

Arthur A. Hyde
Meghan Ormiston
Pamela Hyde

FULL implementation of the NCTM Standards for curriculum, evaluation, and teaching (NCTM 1989, 1991) requires that effective professional development experiences for practicing teachers be created and sustained in schools. Research on the implementation of planned change, dissemination of exemplary practices, diffusion of innovations, and professional development has shown pronounced organizational and cultural differences among schools and districts. This research argues against simplistic prescriptions for professional development that attempt to replicate successes in one setting in another. (See McLaughlin [1987].) No single approach is best.

When providers of professional development design their programs, they should expect to adapt their work to the cultural and organizational features of each setting. Positive collaboration among providers and school people will likely result in what has been called "mutual adaptation." (See McLaughlin and Marsh [1978].) For professional development in mathematics to become a sustained, ongoing part of school culture, vehicles must be created that simultaneously make sense to the school people in their local context and transform their culture.

In this article, we will discuss some aspects of school culture that either facilitate or hinder the changes in teaching and learning promoted in the NCTM *Standards*.

CULTURAL BELIEFS AND NORMS

Much of what anthropologists conceive of as "culture" resides in people's heads: their beliefs, ideas, conceptions, and assumptions about the way things are. These notions get translated into norms for behavior: interpersonal relations, organizational routines, and daily practices. To its participants, culture is subtle and tacit. It is rarely examined.

As participants in American culture, school people have been surrounded by the persistent myths about mathematics teaching and learning of the nineteenth century. Such beliefs as the value of procedural capability regardless of conceptual understanding, the need for complex paper-and-pencil calculations, and the correlation of rapid production of the one right answer to intelligence still abound among teachers and administrators. These beliefs reinforce teaching practices and organizational routines that the NCTM *Standards* urge us to decrease. Because such aspects of schooling have their roots in cultural beliefs and norms, they are remarkably resistant to simplistic and short-term efforts to change.

PERSONAL BREAKTHROUGH EXPERIENCES FOR TEACHERS

Changing the beliefs about mathematics teaching and learning that teachers possess requires giving them powerful experiences in mathematical thinking and conceptual understanding (Hyde 1989). For most elementary school teachers, a very different type of teaching is described in the *Standards* from what they experienced as a student. Merely *telling* teachers what to do differently or *showing* them what to do is insufficient.

Many elementary school teachers have never developed conceptual understanding or rich connections among the mathematical concepts they teach. Many equate memorizing procedures with conceptual understanding. Most elementary school teachers need intense, personal "breakthrough" experiences in mathematics to know and feel the power of conceptual understanding. With such experiences, teachers begin to realize how important understanding is for their students.

Professional development programs should actively engage teachers in doing mathematics, with the leaders facilitating experiences that model what teachers should do with their own pupils. Lampert (1990, p. 34) refers to such activities as creating "a participation structure that redefines the roles and responsibilities of both teacher and students in relation to learning and knowing."

When conducting a workshop in a school with well-established, positive teaching norms about classroom discourse and student collaboration, leaders can relate new conceptions of mathematics teaching to this foundation. For example, some districts have provided teachers with intensive professional development in the writing process, whole language, or cooperative learning. Such experiences foster aspects of school culture that are conducive to students' using manipulatives, writing in their mathematics journals, collaboratively collecting data, and so forth. In contrast, when working with districts that don't have this foundation, leaders must adapt the professional development in order to introduce teachers to student discourse and collaboration.

Openness to Change

Ideally, teachers volunteer for professional development experiences that they believe hold the promise of enhancing their teaching. Such willingness greatly facilitates change. (See Hyde [1992].) Teachers will more likely volunteer for professional development in mathematics if providers have established some rapport and credibility with them. Because of the manifest mathematics anxiety of many elementary school teachers, the affective tone of the providers is critical. They must offer the promise of effective and useful methods in a nonthreatening manner. If a group of teachers participates in a good program, their subsequent enthusiasm will encourage others to participate.

In contrast, when teachers are *required* to attend professional development, their reaction is often resentment and resistance to change, which can readily be translated into closing the classroom door and continuing to teach as usual. A teacher's having "learned" a new approach and successfully demonstrated it in the classroom to a supervisor does not guarantee that it will be habitually used (Joyce and Showers 1983).

Changes in the Classroom

Effective professional development programs should provide methods, materials, and activities that teachers can try out in their classrooms with initially limited risks. The further the activity deviates from a teacher's current practice (or comfort zone [Lovitt et al. 1990]), the more risk in terms of a teacher's perceived ability to manage the activity and control the students' behavior. Most teachers need to attempt minor changes in their practices before making significant movements toward the mathematics teaching envisioned in the *Standards*. Among administrators, however, beliefs vary widely in how changes in teaching should or can occur.

For instance, some administrators believe in an oversimplified training model in which changes in teaching occur when teachers are coached to implement specific activities or materials. Even though elementary school teachers of mathematics can profit from well-crafted materials and activities, professional development is generally conceived of as much broader than training. (See Orlich [1989] and Sparks and Loucks-Horsley [1989].) Understanding why an activity works in terms of its inherent mathematical concepts and thinking, being able to help students make a critical connection in a developmental sequence, and knowing how to modify a model activity are some of the hallmarks of a professional teacher. Providers of professional development may have to work hard at broadening people's conceptions.

BUILDING COLLABORATION AMONG TEACHERS

Professional development activities in mathematics often afford teachers their first opportunity to share mathematical experiences with others. Discourse about mathematical ideas, small-group problem solving, collaborative design of data collection and representation, and the like can be an occasion for powerful, enjoyable, affective experiences. Such collegiality is crucial to establish in the culture of the schools in order to sustain ongoing professional development. (See Lambert [1988], Lieberman [1987], and Little [1987].)

Unfortunately, the norms of isolation and autonomy in most schools undercut collegiality among teachers. Providers of professional development should ask, What are the vehicles for collaboration in the daily lives of the teachers in this school?

Schools are remarkably different in this sociocultural dimension. In some, personal friendships among teachers at various grade levels ensure ongoing dialogue about children and teaching activities. They share ideas and manipulatives and even coteach. In other schools, latent competition among teachers thwarts collaboration. In some schools, the lunchroom tables or the teachers' lounges are centers for talk about teaching. In others, outsiders would hardly know by the teachers' discussions that they were in a school. Building meetings similarly display this range of dialogue about teaching. In some schools, norms exist for teachers to share teaching ideas with one another in a nondidactic, open fashion. In other schools, such sharing would be considered inappropriate and showing off.

If providers were interested in merely *training* teachers in new methods and materials for mathematics teaching, perhaps they could ignore these details. However, if they are committed to stimulating fundamental changes in the culture of mathematics teaching and learning, they must understand how to help teachers learn to share and work with one another. Mathematics workshops can encourage pairs of teachers to design activities together and try them with students, team teaching when feasible. Similarly, as interpersonal relationships develop, teachers may become more willing to demonstrate classroom activities to others in the group and to be observed while teaching, thus taking more risks. When elementary school teachers comfortably teach in front of and with others, the culture of the school has likely begun to change in important ways.

THE ROLE OF PRINCIPALS

Some principals have created extraordinary norms for instructional change. They have introduced ideas into the culture of the school in gradual yet systematic ways: stimulating teachers' interests by sending them to conferences, buying new materials for teachers to try out, making substitutes available so that teachers can observe one another's teaching, and arranging weekly schedules for teachers to have time to plan collaboratively. They have created school cultures that encourage talking about teaching, which allows ideas to "percolate" over time. Teachers get acculturated to new practices, fresh conceptions, and changed norms.

DEVELOPING TEACHER-LEADERS

Sustaining mathematics professional development in schools requires the acknowledgment and support of excellent practicing teachers. Most elementary schools do not have a role designed to assist teachers with mathematics. A special kind of leadership is needed from an "expert peer," a teacher in the school to whom others look for assistance with the new approaches to teaching mathematics. Some excellent elementary school teachers over time have developed the interpersonal relationships within their schools that allow them to provide mutual support to their colleagues. However, the norms of many schools make even this informal role an impossibility. The capability of excellent teachers is often more easily appreciated in other schools or districts.

An important reason for promoting teacher-leaders is their potential for influencing policies and practices within their districts. Some have made presentations to their boards of education about the Standards and how they have implemented them. They have volunteered for mathematics-curriculum committees, thus affecting decisions about texts, manipulatives, and supplementary materials.

Teacher-leaders can work closely with district administrators who have the formal responsibility for decisions about expenditures, resources, the use of inservice time, stipends, and roles that directly affect professional development and curricular change. In this collaboration, administrators' beliefs about the process of curriculum change and professional development are crucial. Driven by budgeting processes, textbook-adoption timetables, and the like, many administrators conceive of no more than one year of change at a time. Teacher-leaders can help broaden the time frame of change and work for a commitment to professional development over three to five years to ensure coherence and continuity.

Providers of professional development should expect to modify their approach to working with districts, administrators, schools, and teachers on the basis of the cultures they find. They can build on those aspects of school culture that will facilitate the changes espoused in the NCTM *Standards.* They also must find ways to promote changes in school culture where needed for professional development to flourish.

REFERENCES

Hyde, Arthur A. "Developing a Willingness to Change." In *Effective Staff Development for School Change,* edited by William Pink and Arthur A. Hyde, pp. 169–88. New York: Ablex, 1992.

———. "Staff Development: Directions and Realities." In *New Directions for Elementary School Mathematics,* 1989 Yearbook of the National Council of Teachers of Mathematics, edited by Paul R. Trafton, pp. 223–33. Reston, Va.: The Council, 1989.

Joyce, Bruce R., and Beverly Showers. *Power in Staff Development through Research on Training.* Alexandria Va.: Association for Supervision and Curriculum Development, 1983.

Lambert, Linda. "Staff Development Redesigned." *Phi Delta Kappan* 69 (May 1988): 665–68.

Lampert, Magdelene. "When the Problem Is Not the Question and the Solution Is Not the Answer: Mathematical Knowing and Teaching." *American Educational Research Journal* 27 (Spring 1990): 29–63.

Lieberman, Ann. "Teacher Leadership." *Teachers College Record* 80 (1987): 400–405.

Little, Judith W. "Teachers as Colleagues." In *The Educator's Handbook: A Research Perspective,* edited by Virginia Richardson-Koehler, pp. 491–518. New York: Longman, 1987.

Lovitt, Charles, Max Stephens, Doug Clarke, and Thomas A. Romberg. "Mathematics Teachers Reconceptualizing Their Roles." In *Teaching and Learning Mathematics in the 1990s,* 1990 Yearbook of the National Council of Teachers of Mathematics, edited by Thomas J. Cooney, pp. 229–36. Reston, Va.: The Council, 1990.

McLaughlin, Milbrey W. "Learning from Experience: Lessons from Policy Implementation." *Educational Evaluation and Policy Analysis* 9 (Summer 1987): 171–78.

McLaughlin, Milbrey W., and David D. Marsh. "Staff Development and School Change." *Teachers College Record* 80 (September 1978): 69–94.

National Council of Teachers of Mathematics. *Curriculum and Evaluation Standards for School Mathematics.* Reston, Va.: The Council, 1989.

―――. *Professional Standards for Teaching Mathematics.* Reston, Va.: The Council, 1991.

Orlich, Donald C. *Staff Development: Enhancing Human Potential.* Boston: Allyn & Bacon, 1989.

Sparks, Dennis, and Susan Loucks-Horsley. "Five Models of Staff Development for Teachers." *Journal of Staff Development* 10 (Fall 1989): 40–57.

6

Diversity, Reform, and Professional Knowledge: The Need for Multicultural Clarity

William F. Tate

IF YOU want a lawyer to represent you in a grievance, would you knowingly hire one whose legal education consisted only of memorizing the laws, codes, and regulations of your state? Most likely not. Imagine having this lawyer prepare a case for you involving several different types of institutions in our society—a government agency, a citizens' group, and a large corporation. It would be important for the lawyer to understand the underlying philosophy of each institution and to integrate this background information with the law to formulate a strong case in your behalf. This requirement implies that a lawyer must have knowledge of the legal system and of the many different types of institutions that interact within this system. The lawyer's reasoning is guided by the context and factors of each case. Merely memorizing the law is inadequate for this profession.

Many of the issues that apply to a lawyer's professional knowledge base apply to the knowledge base of a mathematics teacher, especially in light of the recommendations found in the *Curriculum and Evaluation Standards for School Mathematics* (National Council of Teachers of Mathematics [NCTM] 1989). The *Standards* calls for giving students a mathematics education that prepares them to make decisions about complex technical issues that arise within the social and political context of our democratic society. This approach to mathematics education requires teachers not only to have a thorough knowledge of the possible connections among mathematical topics but also to know how and why mathematics is used in social and political contexts. Thus, teachers must have a knowledge of the kinds of realistic contexts in which mathematics is used. This expanded knowledge base is a requisite to giving students the type of instruction that prepares them to formulate and verify conjectures involving real problems that arise in a democracy.

I wish to thank Kimberly Cash, Maureen Gillette, and Carl Grant for comments on earlier versions of this paper.

We know very little about teachers' understanding of mathematics. Research on the subject-matter knowledge of teachers is a growing field of inquiry in mathematics education (e.g., Ball [1990], Tirosh and Graeber [1990], and Tate [1991]) and teacher education (National Center for Research on Teacher Education 1988). Most of this research can be grouped into two categories. One group of studies examines the influence of teachers' assumptions and beliefs about mathematics on their teaching of the subject (e.g., Ferrini-Mundy [1986] and Thompson [1992]). A second group of investigations focuses on teachers' knowledge of specific mathematics topics (e.g., Ball [1990], Tirosh and Graeber [1990], and Tate [1991]). Yet none of the research on teachers' subject-matter knowledge examines teachers' understanding of mathematics as it applies to society's democratic and economic processes. Thus far, the research on teachers' subject-matter knowledge has been built on the assumption that mathematics is a neutral, objective, abstract, culture-free discipline. Rather than support the vision in the *Curriculum and Evaluation Standards* (NCTM 1989) of a context-based mathematics education, the current state of the research literature may lead those responsible for teacher education to believe that improved educational experiences for teachers should consist only of more decontextualized mathematics subject matter.

Mathematics teacher education built on the principles of a context-free and culturally neutral curriculum is also problematic from an equity perspective. According to Secada (1991, p. 49), an equitable mathematics education would include—

> real contexts that reflect the lived realities of people who are members of equity groups and unless those contexts are as rich in the sorts of mathematics which can be drawn from them as from others, we are likely to stereotype mathematics as knowledge that belongs to a few privileged groups.

Secada's equity argument is grounded in the belief that all children should see themselves as a part of the mathematics curriculum, regardless of their backgrounds. Similarly, Woodson (1990) argued that an equitable education for the African American child would include *more* mathematics education than that received by the white child. He reasoned that the culture of many white children offers them the opportunity to use the abstract arithmetic taught in schools in situations in their homes and business affairs. Woodson characterized the teaching of decontextualized arithmetic as a "foreign method," incapable of empowering African American children because of their lack of opportunity to connect mathematics to their experiences and the experiences they are likely to encounter in society. Woodson's argument can be expanded to include many cultures, races, and socioeconomic groups.

There is a pressing need to prepare mathematics teachers better so they can effectively teach diverse populations (Zeichner 1992; Grant and Secada 1990). A serious question is, What professional knowledge is most appropriate in this endeavor? Unfortunately, this question has received very little attention. Yet the

Professional Standards for Teaching Mathematics (NCTM 1991) recommends that all mathematics teachers know the following:

- How students' linguistic, ethnic, racial, gender, and socioeconomic backgrounds influence their learning of mathematics
- The role of mathematics in society and culture, the contribution of various cultures to the advancement of mathematics, and the relationship of school mathematics to other subjects and realistic applications

The foregoing recommendations call for mathematics teachers to have a multicultural knowledge base. In a review of the literature on multicultural education, Sleeter and Grant (1987, p. 422) found that five approaches were currently in operation in schools. Each of these approaches—"teaching the exceptional and culturally different (TECD)," "human relations," "single group studies," "multicultural education," and "education that is multicultural and social reconstructionist (EMCSR)"—describes a perspective on addressing the needs of culturally diverse students. Two of these multicultural approaches, TECD and EMCSR, will be described. Included in the discussion will be the pedagogical strategies and curricular knowledge base required of a mathematics teacher employing the given approach. These two approaches were selected because they represent two very different ways of providing a multicultural education. "Teaching the exceptionally and culturally different" is an assimilationist approach to multicultural education, whereas the multicultural and social reconstructionist approach seeks to prepare students to analyze social conditions critically (Ladson-Billings 1990).

The purpose of this article is to discuss and describe the role of multiculturalism in meeting the mathematical needs of all children, especially those from diverse populations, so that teachers and teacher educators will have a backdrop for examining and rethinking their professional knowledge base. This analysis will also include a discussion of the potential strengths and problems related to each multicultural approach.

MULTICULTURAL APPROACHES

Teaching the Culturally Different

According to Sleeter and Grant (1987), the goal of the TECD approach to multicultural education is to help the culturally diverse student develop competence in the public culture of the dominant group. Advocates of the TECD approach emphasize building bridges between the various racial and ethnic groups of our society to facilitate individual achievement and social mobility rather than addressing the distribution of power among various groups within the society (Ladson-Billings 1990; Sleeter and Grant 1987). The implications of this approach on the mathematics education of diverse students are twofold.

First, since the ultimate goal of this approach is to have all students assimilate into the American mainstream, culturally different students must acquire the mathematical knowledge that will have the greatest economic return for them and the society. Grant and Sleeter (1989, p. 15) commented on the goal of a culturally different approach to mathematics education:

> Algebra may not seem practical in everyday life, but a knowledge of algebra is useful if one wants to attend college. Teachers should strive to guide exceptional and culturally different learners toward acquiring the knowledge that will help them most later in life. When your students leave high school, they will compete with other students for college entrance, jobs, or scholarships, and teachers need to prepare them for how to compete.

The TECD approach emphasizes the preparation of students to compete within the existing social arrangements. Students are not encouraged to analyze or challenge the underlying principles involved in mathematizing social phenomena that may affect their lives. According to Putnam, Lampert, and Peterson (1990), an important way for students to know mathematics is to understand how mathematics serves as a vehicle to guide the development of social policy. The culturally different approach to multicultural education would not advocate a mathematics education to accomplish this goal.

The culturally different approach can be likened to the "industrial trainer" ideology of mathematics education. Ernest (1991, pp. 147–48) provides insight into the industrial trainer ideology of school mathematics:

> The intrusion of social issues such as multiculturalism, ethnicity, antisexism, antiracism, world-studies, environmental issues, peace and armaments, is rejected outright. Not only are they perceived as irrelevant to mathematics, they undermine British culture.... The consideration of any critical social issue in mathematics teaching is to be strongly opposed. Mathematics ia [*sic*] a value-free tool, so to include such issues usually represents a sinister attempt to undermine its neutrality.

Like the industrial trainer ideology, the culturally different approach to school mathematics rejects the intrusion of social issues, ethnicity, antisexism, and world studies into the curriculum. Instead, the curriculum would consist of basic skills.

Pedagogical knowledge and strategies

Recall that one goal of the TECD approach is to have all students succeed in the traditional algebra curriculum. To accomplish this goal, two pedagogical strategies are seen as relevant. First, a direct approach to teaching is believed to be the most successful method. For the teacher guided by the TECD philosophy, learning is seen as a static process in which the teacher passes information to the student by means of lectures and textbook assignments (see, e.g., Ladson-Billings [1990]). This view is consistent with the industrial trainer ideology of teaching mathematics. Both view teaching as authoritarian. The teacher possesses mathematical knowledge, and his or her role is to impart this knowledge to the student. Teaching is seen as providing opportunities for students to memorize mathematics facts and practice related skills. Proper teaching does not include investigative activities, games, puzzles, surveys, or other activities that deviate from direct instruction on topics that relate to achieving algebra for everyone.

The second pedagogical recommendation derived from the literature on the TECD philosophy involves the role of learning styles and instruction. Before proceeding with more specifics on learning styles, I must caution the reader about the ambiguity in the recommendation on learning styles and in the first pedagogical recommendation of direct instruction. Although both seek to enable students to learn the traditional mathematics curriculum, they diverge on how this learning should be accomplished. The literature on learning styles recommends teaching techniques that take advantage of the student's communication preferences—visual versus auditory perception; structured versus unstructured tasks; and global versus detail orientation (see, e.g., Hale-Benson [1986]). For example, Fennema and Peterson (1987) found that girls learn better in environments that stress cooperation rather than competition. The point of the learning-style recommendations is to help motivate students to learn the traditional curriculum.

Curricular knowledge

Advocates of the culturally different approach see mathematics as a clear set of facts and procedures. These facts and procedures are devoid of a realistic social context. This approach is consistent with a "back to basics" focus. The culturally different student must know the basic facts in order to compete in society. Students' knowledge of these facts must be demonstrated by their performances on traditional evaluative instruments (e.g., textbook work, ditto sheets, and tests). Incorporating calculators and computers into the curriculum is not encouraged. Technological resources are seen as distractions from the goal of developing computational proficiency in the culturally different student.

If problem solving were a part of the curriculum, it would be devoid of contexts related to the lives of the students. The students are given little opportunity to share and integrate their out-of-school mathematical experiences and realities with the mathematics curriculum presented by the teacher.

Strengths and problem areas of TECD

The strength of the TECD approach rests in the belief that teachers should be committed to providing students from diverse groups with a traditional approach to mathematics education. Unfortunately, beyond the efforts of Grant and Sleeter (1989) to expand the array of instructional strategies, the culturally different approach offers teachers very little with respect to pedagogical and curriculum recommendations that adequately address the needs of diverse students. The culturally different approach to multiculturalism does not call for the type of school mathematics being advocated in the reform movement (see, e.g., Lindquist [1989], Fey and Good [1985], and Dossey [1991]). In fact, the culturally different approach is antithetical to recommendations found in the *Curriculum and Evaluation Standards* (NCTM 1989) and the *Professional Standards for Teaching Mathematics* (NCTM 1991). For instance, the culturally different approach—

- relies on an outside authority (teacher) to solve problems;
- teaches isolated mathematics topics out of realistic contexts;
- de-emphasizes the use of calculators and computers;
- uses memorization as a primary means to learn mathematics.

Further, the culturally different approach may promote a "dysconscious" professional knowledge base for the mathematics teacher. King (1991) defined *dysconsciousness* as "an uncritical habit of mind (including perceptions, attitudes, assumptions, and beliefs) that justifies inequity and exploitation by accepting the existing order of things as given" (p. 135). Teachers' dysconsciousness can result in the development of a cultural discontinuity between diverse students and their mathematics education. *Cultural discontinuity* is defined as the incongruence between students' life experiences and the types of school mathematics experiences they are given (see, e.g., Stanic [1991] and Stiff and Harvey [1988]). An incongruence can be curricular or pedagogical. The failure of the TECD approach to address the importance of having diverse students learn mathematics in contexts related to their lives represents a possible curricular incongruence. The emphasis on learning mathematics through memorization rather than exploratory activities exemplifies a potential pedagogical incongruence.

Multicultural and Social-Reconstructionist Education

Mathematical applications are a part of every aspect of our lives. In fact, one of these applications can even predict the chances of our dying. A software program called APACHE III produces a powerful daily statistic: the probability that a patient will die in intensive care or sometime after he leaves the unit (Seligmann and Sulavik 1992). This statistic is calculated on the basis of such variables as a patient's medical history, treatment, and test results.

Should mathematics education provide students an opportunity to discuss and analyze the foregoing situation or other situations that have been mathematized? The answer is yes if the teacher intends to furnish students a mathematics education that is multicultural and social reconstructionist. Grant and Sleeter (1989, p. 212) claimed that one of the goals of this approach was:

to prepare students to be citizens able to actualize the egalitarian ideology that is the cornerstone of our democracy. It teaches students about issues of social equality, fosters an appreciation of America's diverse population, and teaches them political action skills that they may use to deal vigorously with these issues.

According to Grant and Sleeter, the term *social reconstructionist* refers to the philosophical heritage of this approach, which argues for schools to prepare students to participate actively in the democracy by addressing real social problems. An early twentieth-century advocate of the social reconstructionist position was John Dewey. He argued for the implementation of three interconnected sets of beliefs pertaining to this philosophy.

First, Dewey viewed all knowledge as tentative and fallible (Ernest 1991). Dewey's view challenges the epistemological view of mathematics held by advocates of the culturally different approach to multiculturalism, who perceive mathematics as certain and above question. Dewey (1920, p. 137) stated that

> mathematics is often cited as an example of purely normative thinking dependent upon a priori canons and supra-empirical material. But it is hard to see how the student who approaches the matter historically can avoid the conclusion that the status of mathematics is as empirical as that of metallurgy. Men began with counting and measuring things just as they began with pounding and burning them.... The metallurgist who should write on the most highly developed method of dealing with ores would not, in truth, proceed any differently [than a mathematician]. He too selects, refines, and organizes the methods which in the past have been found to yield the maximum of achievement. Logic is a matter of profound human importance precisely because it is empirically founded and experimentally applied. So considered, the problem of logical theory is none other than the problem of the possibility of the development and employment of intelligent method inquiries concerned with deliberate reconstruction of experience.

Second, Dewey argued that education should prepare students for democracy (Ernest 1991). Social reconstructionists believe the school is one of the primary places where children learn how to collectively make an impact on social institutions. Advocates of this philosophy argue that schools should teach and model democratic living. In this way, students will be prepared for the political aspects of a democracy.

Third, Dewey proposed that education should be built on the child's interests and experiences (Ernest 1991). Education becomes meaningful and real to students when it is connected to them personally, as opposed to using materials that may be abstract and unrelated to a child's lived reality. It appears that Woodson (1990) and Dewey (1920) recognized that premising a mathematics education on students' learning abstractions was truly a foreign method. Linking education to the child's interest is a conscious effort to avoid this foreign method of teaching mathematics and focuses on making schooling multicultural.

Another way in which the EMCSR approach addresses diversity is in promoting critical thinking by situating students in realistic problem contexts found in our multicultural society and having them analyze and solve the problems. Grant and Sleeter (1989, p. 212) elaborate:

> The phrase *education that is multicultural* refers to the kinds of social issues of greatest concern in a democracy and to the belief that the entire school experience should be reoriented to address these concerns. The issues include social inequality based on race, class, disability, and gender, and the primary concern is whether we are practicing our democratic and egalitarian ideals if certain groups of people continue to oppress and control other groups.

How would a student approach a mathematics problem from a multicultural and social-reconstructionist perspective? First, the problem would in all likelihood have a context derived from a realistic situation. The case of the APACHE III computer program illustrates this point. A student analyzing this mathematized situation must attack the problem like a lawyer. The student should be prepared to establish the underlying premises that lead to and guide the implementation of the mathematical model. According to Schlag and Skover (1986), one tactic used to attack a legal premise is to scrutinize the definition of terms. This tactic is appropriate for analyzing the APACHE III program. Like a lawyer, the student would examine the definition of terms, or in the case of APACHE III, the variables and categories used as inputs into the system. The student must decide if the definition of each variable or category is sufficiently precise to be used in this application. Also, the possibility of the variables' having more than one meaning must be investigated. Are any of the variables or categories value laden or politically charged? For instance, does the system place a patient in an at-risk category on the basis of the patient's race, class, gender, or socioeconomic status? If so, how does the program influence the patient's chances for life support? Does the system eliminate those patients who may not be able to afford the treatment? In fact, medical insurers are requesting that hospitals give them information obtained by the APACHE III system (Seligmann and Sulavik 1992). The possibility is real that insurance companies may set rules and thresholds based on a patient's APACHE III scores. Thus, the variables and categories of the system have consequences for both the patient's insurance benefits and the health services he or she receives.

Analyzing the APACHE III system from a multicultural and social-reconstructionist perspective supplies the student with a lens through which to observe the connection between mathematics—in this example, data analysis—and the goals of health-care institutions and insurance companies. For example, the student investigating this situation might find that health-care institutions have two goals. One goal might be to give each intensive care patient the best health care possible. A second goal could be to optimize the use of hospital facilities. Similarly, the insurance company would want to optimize their investment by providing coverage to those least likely to need it. Finding out how these institutional goals might affect the lives of various groups within the democracy is an important part of the solution process for students working from a multicultural, social-reconstructionist perspective in their mathematics education.

Like a lawyer's, the student's reasoning is guided by the context and factors of the situation. Merely memorizing mathematics facts is inadequate for this approach to multiculturalism. One theory of learning, situated cognition, reaffirms the belief that situations coproduce knowledge through activity, as explained by Brown, Collins, and Duguid (1989, p. 33):

> All knowledge is, we believe, like language. Its constituent parts index the world and so are inextricably a product of the activity and situations in which they are produced.... This would also appear to be true of apparently well-defined, abstract technical concepts ... part of their meaning is always inherited from the context of use.

The goal of the multicultural, social-reconstructionist approach is to prepare students for critical thinking in a democracy. With this goal in mind, the situated-cognition theory of learning would suggest that students ought to be presented problem contexts that involve mathematics and that are like those found in authentic activities of the democracy. The EMCSR approach advocates the learning of mathematical concepts that are derived from realistic social situations.

Pedagogical knowledge and strategies

A teacher working from a multicultural, social-reconstructionist approach attempts to create a learning environment that is "as democratic and open as the power asymmetries of the classroom allow, but with explicit recognition of this asymmetry" (Ernest 1991, p. 209). According to Ladson-Billings (1990), a teacher working from this perspective encourages a community of learners. Similarly, D'Ambrosio (1990) argued that instruction should generate "the dynamics for interactive behavior in situations proposed by the environment" (p. 23). Thus the pedagogical strategies required of this approach must facilitate genuine discussion between the students and the teacher. To achieve the goals of this approach, Ernest (1991) suggested that teachers organize their instruction to include the following:

- Cooperative group work, projects, and problem solving to promote engagement and mastery
- Autonomous projects, problem posing, and investigative work to afford students opportunities for self-directed and personally relevant activities

Assessment would take a variety of forms, including portfolios, projects, and traditional examinations. The assessment tasks would depend on, and be consistent with, the original problem context and setting. The emphasis is on the student's ability to illustrate critical thinking about mathematical concepts and about the role of these concepts in making decisions in society. Assessment tasks must furnish evidence that the students have resolved, or understand the path to resolving, any findings they deem unsatisfactory from their perspective as citizens in the democracy.

Curricular knowledge

An EMCSR teacher views school mathematics as a social construction. Mathematics is developed as a result of "human creation and decision-making, and connected with other realms of knowledge, culture and social life" (Ernest 1991, p. 207). Thus, this approach requires the mathematics teacher to keep abreast of realistic situations that use mathematics. The teacher should also have access to resources that use realistic contexts to facilitate students' investigations and projects. The EMCSR teacher views the curriculum as the body of relevant information that provides students with a context for discussing and analyzing mathematics in realistic situations in the society and in their lives.

For example, Frankenstein (1990) used data from such sources as the Census Bureau and the Department of Commerce as a way to situate students in real contexts and to get them to resolve basic statistical questions. Part of the solution process was to analyze and make inferences from statistical abstracts. Students were encouraged to draw on several sources to justify their conjectures. Further, her students discussed why some data sources use race, class, and gender as categories and others do not. I contend that this type of discussion is important if schools are to prepare students to use mathematics in the democracy.

Strengths and a problem of the EMCSR approach

The multicultural and social-reconstructionist approach to mathematics education has two major strengths. First, it represents an effort to use mathematics to promote equity and fairness in the democracy. This effort includes preparing students for competing in both the economy and the democracy. More specifically, students are taught to think critically about mathematics and its use in society. The EMCSR approach looks to develop producers and critical analyzers of mathematics. In contrast, the TECD approach focuses on developing mathematical consumers.

Second, the EMCSR approach is consistent with contemporary visions of mathematics education. For instance, both the NCTM *Curriculum and Evaluation Standards* (1989) and the *Professional Standards for Teaching Mathematics* (1991) call for teachers to provide instruction that encourages students to make connections among mathematics, other disciplines, social realities, and the students' lived realities. An education that is multicultural and social reconstructionist would offer students these opportunities.

According to Ernest (1991), a problem in an approach like EMCSR is the conflict that can arise in classroom settings. Interestingly, the very nature of mathematics and its advancement is based on conflict (Apple 1979). Mathematicians resolve their disputes by arguing for their conjectures. Colleagues examine their arguments for fallacious reasoning and possible counterexamples. Similarly, conflict is a natural product of democracy. In theory, each citizen looks to have his position on an issue recognized as being sound and logical. Conflicts arise when many diverse opinions and positions must be addressed. Thus, preparing students for conflict actually prepares them for the realities of both mathematics and the democracy. I contend that this "problem" is actually a strength of the approach.

FINAL NOTE

Multiculturalism is an important aspect of the professional knowledge base of mathematics teachers. Yet it is often misunderstood. It is important for teachers and teacher educators to be attuned to the epistemological underpinnings of the various approaches to multiculturalism and how these approaches can inform and guide their teaching of mathematics.

REFERENCES

Apple, Michael W. *Ideology and Curriculum*. Boston: Routledge, 1979.

Ball, Deborah Loewenberg. "Prospective Elementary and Secondary Teachers' Understanding of Division." *Journal for Research in Mathematics Education* 21 (March 1990): 132–44.

Brown, John S., Allan Collins, and Paul Duguid. "Situated Cognition and the Culture of Learning." *Educational Researcher* 18 (January-February 1989): 32–42.

D'Ambrosio, Ubiratan. "The Role of Mathematics Education in Building a Democratic and Just Society." *For the Learning of Mathematics* 10 (November 1990): 20–23.

Dewey, John. *Reconstruction in Philosophy*. New York: Holt, 1920.

Dossey, John A. "Discrete Mathematics: The Math for Our Times." In *Discrete Mathematics across the Curriculum: K–12,* 1991 Yearbook of the National Council of Teachers of Mathematics, edited by Margaret J. Kenney, pp. 1–9. Reston, Va.: The Council, 1991.

Ernest, Paul. *The Philosophy of Mathematics Education*. Bristol, Pa.: Falmer Press, 1991.

Fennema, Elizabeth, and Penelope Peterson. "Effective Teaching for Girls or Boys: The Same or Different?" In *Talks to Teachers,* edited by David C. Berliner and Barak V. Rosenshine, pp. 111–25. New York: Random House, 1987.

Ferrini-Mundy, Joan. "Mathematics Teachers' Attitudes and Beliefs: Implications for Inservice Education." Paper presented at the annual meeting of the American Educational Research Association, April 1986, San Francisco.

Fey, James T., and Richard A. Good. "Rethinking the Sequence and Priorities of High School Mathematics Curricula." In *The Secondary School Mathematics Curriculum,* 1985 Yearbook of the National Council of Teachers of Mathematics, edited by Christian R. Hirsch, pp. 43–52. Reston, Va.: The Council, 1985.

Frankenstein, Marilyn. "Incorporating Race, Gender, and Class Issues into a Critical Mathematical Literacy Curriculum." *Journal of Negro Education* 59 (Summer 1990): 336–51.

Grant, Carl A., and Walter G. Secada. "Preparing Teachers for Diversity." In *Handbook of Research on Teacher Education,* edited by W. Robert Houston, pp. 403–22. New York: Macmillan Publishing Co., 1990.

Grant, Carl A., and Christine E. Sleeter. *Turning on Learning: Five Approaches for Multicultural Teaching Plans for Race, Class, Gender, and Disability*. Columbus, Ohio: Merrill Publishing Co., 1989.

Hale-Benson, Janice E. *Black Children: Their Roots, Culture, and Learning Style*. Baltimore, Md.: Johns Hopkins University Press, 1986.

King, Joyce K. "Dysconscious Racism: Ideology, Identity, and the Miseducation of Teachers." *Journal of Negro Education* 60 (Spring 1991): 133–46.

Ladson-Billings, Gloria. "Like Lightning in a Bottle: Attempting to Capture the Pedagogical Excellence of Successful Teachers of Black Students." *Qualitative Studies in Education* 3 (1990): 335–44.

Lindquist, Mary Montgomery. "It's Time to Change." In *New Directions for Elementary School Mathematics,* 1989 Yearbook of the National Council of Teachers of Mathematics, edited by Paul R. Trafton, pp. 1–13. Reston, Va.: The Council, 1989.

National Center for Research on Teacher Education. "Teacher Education and Learning to Teach: A Research Agenda." *Journal of Teacher Education* 39 (November-December 1988): 27–32.

National Council of Teachers of Mathematics. *Curriculum and Evaluation Standards for School Mathematics.* Reston, Va.: The Council, 1989.

———. *Professional Standards for Teaching Mathematics.* Reston, Va.: The Council, 1991.

Putnam, Ralph T., Magdalene Lampert, and Penelope L. Peterson. "Alternative Perspectives on Knowing Mathematics in Elementary Schools." In *Review of Research in Education,* edited by Courtney B. Cazden, pp. 57–152. Washington, D.C.: American Educational Research Association, 1990.

Schlag, Pierre, and David Skover. *Tactics of Legal Reasoning.* Durham, N.C.: Carolina Academic Press, 1986.

Secada, Walter G. "Agenda Setting, Enlightened Self-Interest, and Equity in Mathematics Education." *Peabody Journal of Education* 66 (Winter 1991): 22–56.

Seligmann, Jean, and Chris Sulavik. "Software for Hard Issues." *Newsweek,* 27 April 1992, p. 55.

Sleeter, Christine E., and Carl A. Grant. "An Analysis of Multicultural Education in the United States." *Harvard Educational Review* 57 (November 1987): 421–44.

Stanic, George M. A. "Social Inequality, Cultural Discontinuity, and Equity in School Mathematics." *Peabody Journal of Education* 66 (Winter 1991): 57–71.

Stiff, Lee V., and William B. Harvey. "On the Education of Black Children in Mathematics." *Journal of Black Studies* 19 (December 1988): 190–203.

Tate, William F. "An Evaluation of the Function Knowledge of a Selected Group of Prospective Secondary Mathematics Teachers." Doctoral dissertation, University of Maryland, 1991.

Tirosh, Dina, and Anna O. Graeber. "Evoking Cognitive Conflict to Explore Preservice Teachers' Thinking about Division." *Journal for Research in Mathematics Education* 21 (March 1990): 98–108.

Thompson, Alba G. "Teachers' Beliefs and Conceptions: A Synthesis of the Research." In *Handbook of Research on Mathematics Teaching and Learning,* edited by Douglas A. Grouws, pp. 127–46. New York: Macmillan Publishing Co., 1992.

Woodson, Carter G. *Miseducation of the Negro.* 1933. Reprint. Trenton, N.J.: Africa World Press, 1990.

Zeichner, Kenneth. *Educating Teachers for Cultural Diversity.* East Lansing, Mich.: Michigan State University, National Center for Research on Teacher Learning, 1992.

7

Changing Mathematics Teaching Means Changing Ourselves: Implications for Professional Development

Julian Weissglass

IT IS widely acknowledged that the vision of mathematics classrooms presented by NCTM's two *Standards* documents is radically different from current practice and that implementing this vision will require considerable re-education of current teachers. Although adequate implementation will require policy and structural changes, the nature of the professional development in which teachers participate will largely determine the extent of the change in students' classroom experiences. The failure of past reform efforts to achieve sustainable change indicates that a significant departure from traditional professional development is needed.

In this article I shall analyze the difficulty of change, offer a theoretical perspective on professional development, and describe some experiences using this perspective in professional and leadership development efforts. Finally, by making six recommendations, I shall raise some issues for the mathematics education community to consider as it continues its implementation efforts.

In essence, this article proposes a change in the very way we think about and approach educational reform. It suggests that a major paradigm shift is required. In the old paradigm, educational change is mandated by policy makers or curriculum reformers and delivered by telling teachers about the desired reforms. In the new paradigm, change occurs as the result of complicated interactions among people (teachers, parents, students, university faculty, administrators, policy makers) with their own beliefs and values. At the heart of the process are classroom teachers—recognized and respected as complex human beings, who—along with knowledge—have feelings, beliefs, and values that must be addressed if change is to occur.

THE DIFFICULTY OF CHANGE

Early in this century Maria Montessori (1967) captured the problem when she pointed out that "nothing is more difficult for a teacher than to give up her old habits and prejudices." It was Sarason (1971), however, who linked educational change to cultural change, thereby providing an explanation of why so many attempts to change education have failed. The idea that schools have cultures has profound implications for educational reformers. It requires first of all that we define culture. For the purposes of reflecting on classroom practices in mathematics education, I find the following definition of culture useful: *Culture is the attitudes, beliefs, values, and practices shared by a community of people which they often do not state or question and of which they may not be consciously aware.* That, or a similar definition, leads us to examine current practices, beliefs, and values in the mathematics classroom.

Current practices often have a teacher in front of the classroom with perhaps thirty students sitting at desks. Most class time is spent listening to the teacher or doing computations or algebraic manipulations. Students work individually and at certain times (e.g., exam time) very anxiously. Most of the mathematics introduced at the precollege level was developed between 1475 and 1700 (except for geometry, which was developed before 250 B.C.). Students mainly memorize facts and procedures rather than think mathematically and communicate their thinking. The percentage of female, African-American, and Latino/Latina students in mathematics classes decreases with age. Teachers rarely talk meaningfully to one another about their teaching, their beliefs about teaching and learning, their feelings about their work and the proposed changes, or issues related to racial and class bias.

When I asked a group of teachers to speculate on the beliefs and values underlying these practices, they came up with a long list (Weissglass 1992), some of which follow:

- You must master the content before you can use your brain to think mathematically.
- People learn mathematics by listening to someone talk about it and from doing homework problems.
- Competition is necessary to stimulate learning.
- Practice makes perfect.
- Mathematics is too difficult to understand on one's own—students need to be told.
- Making mistakes is a sign of weakness.
- Mathematics is best developed linearly.
- It is OK not to be good at math.
- Students are incapable of deciding what to learn (such decisions are made by the teacher, district, or state).
- Preparation for success in calculus is the goal.
- It is cheating to get help from another person.
- It is rude to challenge adults.
- We have the ability to accurately measure what students understand.

• Feelings are not part of the academic environment.
• The system is OK (after all, I succeeded).

Teachers whose practices are based on beliefs such as these will be resistant to change unless their beliefs change, which will require personal transformations. Mathematics educators have been slow, however, to recognize that personal transformation leading to changed pedagogy often entails emotional struggle. It does not simply happen from a desire or a request to change.

A MODEL FOR ADDRESSING TEACHERS' FEELINGS AND BELIEFS

Most professional development efforts in mathematics education focus on teachers' knowledge, understandings, or behavior, rarely addressing the emotional aspects of their lives. Although the collegiality and affection that often develop at an institute are marveled at, they are not considered as making a decisive contribution to learning and professional growth. The development of personal relationships is often left to meals or breaks instead of being structured into the professional activities. Feelings of anger, apathy, hopelessness, sadness, or fear are rarely addressed purposefully in professional development. It is unwise and counterproductive, however, for reformers to ignore the fact that the current effort at reform is occurring at a time when schools are dealing with the effects on children of divorce, single-parent families, alcoholism, homelessness, violence, racial prejudice, sexual and physical abuse, and the widespread availability of drugs. In many schools teachers are faced with increased class sizes and clerical responsibilities, declining resources, and pressure to raise standardized test scores. These conditions can stimulate a variety of feelings in teachers. The result is that teachers and principals may be experiencing high levels of stress that affect their ability to even consider—much less implement or sustain—the processes of change. (One definition of teacher stress is "the experience by a teacher of unpleasant emotions such as tension, frustration, anxiety, anger, depression, resulting from aspects of his work as a teacher" [Kyriacou 1987, p. 146].) In addition, many teachers have been inadequately prepared to teach mathematics and are scared of it. Their conception of the nature of mathematics may differ substantially from that of the reformers. Requests to change their teaching methods can increase their anxiety. Although increasing teachers' understanding of mathematics and mathematics pedagogy is an important part of any professional development effort, it would be wise under the current circumstances to also include methods that address feelings and beliefs.

The model (Weissglass 1991) in figure 7.1 is based on the assumption that educational change requires personal transformation and improved collegial relationships. It has four components, viewed as the vertices of a tetrahedron. The model implies that no one component is more important than any other and that each component is essential to the change process:

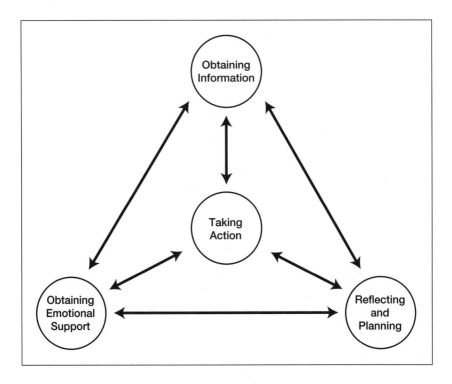

Fig. 7.1

• *Obtaining information.* We need to continue to learn about mathematics and pedagogy; about how to develop better relationships with students, parents, and administrators; and about the effects of bias and mistreatment on students' learning. We need to learn about the reality of the school experience of people not from the dominant culture and of people who have not been successful in school.

• *Reflecting and planning.* We need to reflect on our teaching and leadership, our beliefs, and our theories. We need to plan for instruction and develop strategies for school change. It is useful to set both long-range and short-range goals. Without long-range goals, we may meet new demands but not use our time and energy most effectively. Without short-range goals, we may stay preoccupied with the "vision" without taking the "small steps" necessary to achieve it.

• *Obtaining emotional support.* Unresolved feelings from the past often interfere with our present functioning. They may hinder our learning, confuse our thinking, interfere with our planning, inhibit us from taking action, or cause us to act on the basis of negative feelings rather than our best thinking. Whether negative feelings come from the past or are elicited by the present situation, obtaining emotional support can help us deal with new situations constructively and creatively.

• *Taking action.* The other three vertices culminate in the changes that teachers make in their classrooms and in their interactions with administrators, parents, and one another.

It should be clear that the components interact. A planning activity may indicate a need to learn more mathematics or to work through some feelings in order to act on a decision. Working through feelings leads to increased understanding of new information and perhaps to making decisions to act differently. Acting differently will often stimulate feelings. Expressing such feelings may free up areas to think about and require new reflection and planning. Getting new information may also require reflection and stimulate feelings. For example, learning about the low success rate of students of color may cause you to reflect on what you do in your class. It may also generate strong feelings about a social system that perpetuates inequity.

ADDRESSING THE NEGLECTED VERTICES: CONSTRUCTIVIST LISTENING

The components of the tetrahedral model that are most neglected in the lives of teachers are "Reflecting and Planning" and "Obtaining Emotional Support." The recent interest in teacher reflection (Hopfengardner and Leahy 1988; Schon 1983) is encouraging. Less attention has been given to "Obtaining Emotional Support." I have discussed the reasons that educators neglect emotions (Weissglass 1990). It is hopeful that some segments of society are realizing the importance of emotional support. A *Newsweek* magazine cover story on support groups estimated that 15 million Americans attend support groups each week (Leerhsen 1990). No doubt some of these are educators. They get support in a variety of areas: health, diet, troubling life situations. Not many teachers, however, are getting organized emotional support for teaching, one of the most stressful jobs in this society.

My work with teachers addresses these two components by creating opportunities to be "listened to"—attentively and with care. Listening, however, means different things to different people. I have described elsewhere (Weissglass 1990) several different forms of listening:

• *Active listening,* in which listeners reflect back their interpretation of what the talker is saying

• *Passive listening,* in which the listener doesn't say anything but indicates interest and attention by maintaining eye contact and periodically nodding or smiling

• *Inattentive listening,* where the sounds enter someone's ears but there is little or no attempt to comprehend or respond

• *Pretend listening,* where a person maintains an interested facial expression but is actually thinking about something else

- *Conversational listening,* where the roles of talker and listener alternate, often frequently
- *Argumentative listening,* similar to *conversational listening* but more passionate, with each person looking for flaws in the other's argument
- *Informational listening,* where someone wants information that someone else possesses and attempts to make sense of the information that is received

I encourage teachers to learn and use a different form of listening, which I call *constructivist,* or supportive, listening. Its goals are to encourage the talker to reflect on the meaning of events and ideas, to express and work through feelings that are interfering with clear thinking, to construct new meanings, to make decisions. I use the term *constructivist* because "from the constructivist perspective, learning is the product of self-organization" (von Glasersfeld 1989, p. 136), and the goal of the process is to facilitate this self-organization. Constructivist listening acknowledges a strong relationship between human cognition and emotion. It assumes that teachers must construct their own understanding of teaching. The constructivist listener aims to enable the talker to express feelings, construct personal understandings, and use his or her full intelligence to respond creatively to situations rather than rely on habit or ineffective coping strategies. It is *not* important in constructivist listening for the listener to understand completely what the talker is expressing. It *is* important that the listener communicate interest, caring and acceptance, and through the asking of thoughtful questions help the talker express and explore thoughts and feelings.

In constructivist listening, interpretations by the listener are avoided because they usually interfere with the talker's fully exploring the thought or feeling, expressing emotion, and developing understanding. Constructivist listening is not passive listening, however. The constructivist listener purposefully thinks about the talker and encourages deep exploration of thoughts and feelings by asking appropriate questions that focus the talker's attention. The listener also provides reassurance that it is permissible to delve into murky areas and to express feelings. Constructivist listening differs significantly from the conversational or argumentative listening engaged in by most educators where advice, opinions, and interruptions are commonplace. Even educators who are good listeners—and there are many—rarely listen attentively to one another's feelings or expression of emotions. If a person starts to cry about a painful experience, for example, someone may say something like "Oh, well, try to not let it get to you" or "No use crying over spilled milk." More frequently listeners will communicate their discomfort with a facial expression or body language and the person will regain control and apologize for being so emotional.

APPLICATIONS IN PROFESSIONAL DEVELOPMENT

It takes some time and effort to learn constructivist listening. In my first contact with teachers participating in professional development I ask them to pair up, introduce themselves, and tell each other their goals for the event. I use informal pairings (called dyads) frequently, without introducing any formal structure or theory. The questions the participants are asked to talk about in dyads depend on what I hope to accomplish. For example, if I wish to have educators understand, on a personal level, the difference between *Standards*-like mathematics classrooms and traditional ones, I follow a hands-on mathematics activity by asking them to pair up and talk about how the experience compares with the way they learned mathematics in school. If my goal is to have them appreciate the mathematical concepts in an open-ended investigation, I will ask, "What mathematical concepts were contained in that activity?" Often the dyad is followed by a discussion in small or large groups. For example, in order to help teachers reflect on their assumptions about learning and teaching and to articulate their own theory of learning, I ask them to pair up and relate a memory of a good or exciting mathematical learning experience. They then discuss the characteristics of that experience. Throughout I remind them that it is fine to express their doubts and frustrations as well as their enthusiasm. This is in accord with the advice of Marris (1986, p. 155):

> No one can resolve the crisis of reintegration on behalf of another. Every attempt to preempt conflict, argument, protest by rational planning, can only be abortive: however reasonable the proposed changes, the process of implementing them must still allow the impulse of rejection to play itself out. When those who have power to manipulate changes act as if they have only to explain, and, when their explanations are not at once accepted, shrug off opposition as ignorance or prejudice, they express a profound contempt for the meaning of lives other than their own.

Although constructivist listening is definitely intended to help teachers with the "crisis of reintegration," it is *not* intended to replace other methods that educators use to learn, communicate, or solve problems. It does not fill the need for new information or dialog, nor does it eliminate the need for changes in policies or practices that impede reform. Constructivist listening is intended as an *additional* tool for assisting professional and personal growth. Two further examples will indicate how it can be integrated into a professional development event:

1. Figure 7.2 is a handout from a ninety-minute segment given at a summer institute. The purpose was to (*a*) increase teachers' understanding of symmetry and geometrical transformations, (*b*) convince them (especially the secondary school teachers) that manipulative materials can be used to investigate mathematics at many levels (the ultimate classification of strip patterns relies on advanced group theory), and (*c*) raise teachers' awareness of cultural issues in mathematics education— including ethnomathematics (D'Ambrosio 1988).

Dyad: Relate a memory of seeing something symmetrical or appreciating symmetry. What does symmetry mean to you?

People throughout history have been fascinated with symmetry. They have used symmetry in decorating buildings, clothing, tapestries, and pottery. Figure *a* (reprinted with permission from the article "The Geometry of African Art I. Bakuba Art," by Donald Crowe, which appeared in *Journal of Geometry* [vol. 1, no. 2, p. 176]) is from Zaire, and figure *b* (reprinted with permission from Peter S. Stevens, *Handbook of Regular Patterns* [MIT Press]) is Celtic in origin.

a   b

These patterns provide a possibility for engaging students in mathematics that is close to their culture. For example, the San Ildefonso Pueblo potters decorated their pottery with a variety of strip patterns. Some of them are shown in figure *c* below (reprinted with permission from *The Pottery of San Ildefonso Pueblo,* by Kenneth Chapman [University of New Mexico Press, Plate IV, p. 59]).

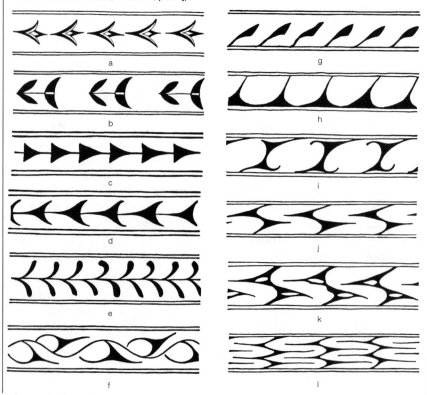

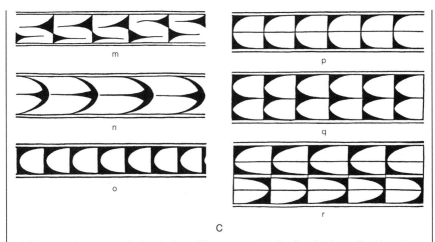

c

1. Discuss what you see in the designs. Why do you think the San Ildefonso Pueblo potters used these designs? Were they doing mathematics?
2. Each person in your group create a strip design with the pattern blocks.
3. Compare and contrast your designs. How are they similar? How are they different?
4. How might you classify the different types of strip patterns?
5. Discuss again: Were the San Ildefonso potters doing mathematics?

Questions for discussion:
• How might you adapt and extend the ideas in this activity at your grade level?
• Suppose you were criticized by an administrator for not following your district's scope and sequence. How could you use Standard 1 from the NCTM *Professional Standards for Teaching Mathematics* (1991) to justify students' doing this activity?
• Prepare a report for the entire group.

Fig. 7.2

2. At a daylong in-service program for primary school teachers on the relationship of play and mathematics learning, teachers built towers from straws and Plasticine, explored manipulatives, and discussed the mathematical concepts involved in the activities and how to adapt them for their grade level. They formed dyads and discussion groups to reflect on the following questions: How did adults react to your play as a child? Are you comfortable or uncomfortable now with children's play? What makes you feel that way? How do you feel if children direct the play in a different direction from what you intended it to be? How do you feel when they don't follow the rules? How do you feel when children get silly?

RECOMMENDATIONS

Although new information is important, furnishing it is not sufficient to overcome personal resistance to change, the tenacity of the existing culture, and the working conditions of educators. Providing opportunities for reflection and planning and for expressing and working through feelings about mathematics learning and teaching will increase the likelihood of teachers developing new understandings, challenging beliefs and assumptions, and changing rigid and unproductive practices. In particular, I recommend that educators leading professional development in mathematics education deliberately take the following six actions:

1. *Combat the isolation under which educators work.* Integrating support groups and dyads into professional development activities will enable educators to develop stronger bonds with their colleagues. These improved relationships will provide educators the safety to discuss meaningfully the issues that they face. As one teacher put it, "Support groups are perhaps the most outstanding innovation educationally. [They] allow teachers to become friends and supporters at a depth not approached with any other group that I have known."

2. *Help educators improve their listening skills.* Educators will generally not become better listeners by being told to listen better or given lectures about it. They become better listeners through practice, reflection, discussion, and the experience of being listened to well. Dyads are good practice, but reflection and the sharing of experiences and insights are necessary. Some questions for dyads and discussion that will help educators learn to listen better follow:

   • Describe a time when someone listened to you attentively without analyzing or judging what you said. How did you feel?
   • When was a time you listened well to someone?
   • Describe a time when you started to talk about an experience, problem, or feeling and someone interrupted or did not pay attention. How did you feel?
   • When do you find it difficult to listen attentively without interrupting?

   Another valuable activity is to review (after participating in a dyad) where you were tempted to interrupt (or when you actually did). If you discuss this in a group, do it without divulging what your partner was talking about, or get his or her permission first.

3. *Provide opportunities for educators to express their feelings about changes being proposed, about issues in education, and about their own past experiences as learners.* To increase participants' acceptance of emotional expression, it is helpful to address the issue directly. Passages from Weissglass (1990) or Weissglass and Weissglass (1987) are useful for information and as a stimulus for discussion. These questions will help educators be more comfortable with emotional expression:

   • How did your family treat feelings? What happened when you were upset and expressed your emotions? How did people react to you if you expressed good feelings?

- How do you feel when young children cry or get angry? What might cause these feelings?
- What causes you to feel bad about yourself?
- How do you handle good feelings? Do you talk about your successes with pride?

4. *Enable educators to address controversial issues on a personal as well as intellectual level.* It is crucial that professional development activities address controversial issues (e.g., tracking, assessment, racial bias). Change will occur, however, only if educators can listen to one another (and community members) respectfully, work through their feelings about the issues, and make decisions for themselves about new approaches.

5. *Provide the opportunity, whenever information is presented, for each person to reflect on its meaning for himself or herself and to plan on how to use that information.* It is unrealistic to expect teachers to implement new ideas in their classrooms unless they have time for reflection and planning. At the very least, for every fifteen minutes of lecture give participants two to three minutes each in a dyad to talk about their understanding of what has been said. For each hour of presentation include at least one-half hour for processing, reflection, and discussion on implementation.

6. *Establish and sustain mutual support networks.* Teachers need to meet regularly with the same people in order to develop the trust necessary to talk meaningfully about important issues. Incorporate support-group meetings into all activities and design staff development so that people can attend regularly. Avoid single sessions.

As support groups develop, it will be important to clarify the difference among discussion groups, support groups, and "action groups." The purpose of support groups is to develop trust among educators, to help each participant personally think through issues, work through feelings, and set goals. These groups are for well-functioning people who want support for making changes. Support groups are not psychotherapy, although they often produce personal growth. (See IMEDC Team [1993] for more discussion of this issue.) They are also not meant to replace discussion or action groups. The purpose of discussion groups is to exchange information, thoughts, and opinions about issues in education and teaching practices. Action groups may go under the name of task force, committee, council, or caucus and may be sanctioned or unsanctioned. Their goal is to take action on procedures or policy. For example, action groups may develop policy for a school or district or challenge existing policy. Although dyads may be useful in action groups for venting feelings and clarifying thinking, educators should not confuse action groups and support groups. Principals in particular need to understand that support groups exist to enable teachers to exchange support for their personal and professional growth, not to challenge policy.

CONCLUSION

There are two unavoidable realities about educational change: (1) The most important aspects of how the classroom operates are under the control of the classroom teacher, and (2) teachers have feelings about what they are doing and what they are being asked to do.

Policymakers and reformers may wish it were different, but it is not. Although we can challenge teachers with different visions of mathematics pedagogy, new curriculum, and research results, we must do more. We must give them opportunities to develop trusting collegial relationships so that they can reflect on their beliefs, construct their own understanding of the proposed changes, work through feelings that may inhibit their ability to change, and make their own decisions about how to respond—both to their students and to the suggestions from the reform movement.

REFERENCES

D'Ambrosio, Ubiratan. "Ethnomathematics and Its Place in the History and Pedagogy of Mathematics." *For the Learning of Mathematics* 5 (1988): 44–48.

Hopfengardner, Jerrold D., and Peggy E. Leahy. "Providing Collegial Support for Experienced Teachers." *Journal of Staff Development* 9 (1988): 48–51.

IMEDC (Improving Mathematics in Diverse Classrooms) Team. *Reflections on Educational Change Support Groups.* Santa Barbara, Calif.: Center for Educational Change in Mathematics and Science, University of California—Santa Barbara, 1993.

Kyriacou, Chris. "Teachers' Stress and Burnout: An International Review." *Educational Research* 29 (1987): 146–51.

Leerhsen, Charles, with others. "Unite and Conquer." *Newsweek,* 5 February 1990, pp. 50–55.

Marris, Peter. *Loss and Change.* Rev. ed. London: Routledge & Kegan Paul, 1986.

Montessori, Maria. *The Discovery of the Child.* New York: Ballantine Books, 1967.

Sarason, Seymour B. *The Culture of the School and the Problem of Change.* Boston: Allyn & Bacon, 1971.

Schon, Donald. *The Reflective Practitioner.* New York: Basic Books, 1983.

von Glasersfeld, Ernst. "Cognition, Construction of Knowledge, and Teaching." *Synthese* 80 (July 1989): 121–40.

Weissglass, Julian. "Changing the Culture of Mathematics Instruction." *Journal of Mathematical Behavior* 11 (1992): 195–203.

———. "Constructivist Listening for Empowerment and Change." *Educational Forum* 54 (1990): 351–70.

———. "Teachers Have Feelings: What Can We Do about It?" *Journal of Staff Development* 12 (1991): 28–33.

Weissglass, Julian, and Theresa Liebscher Weissglass. *Learning, Feelings and Educational Change.* Santa Barbara, Calif.: Kimberly Press, 1987.

8

Calculational and Conceptual Orientations in Teaching Mathematics

Alba G. Thompson
Randolph A. Philipp
Patrick W. Thompson
Barbara A. Boyd

How mathematics curriculum reform is implemented in the classroom depends largely on teachers' images of the mathematics they are teaching (Bauersfeld 1980; Cooney 1985; Thompson 1984). From our close collaboration with middle school mathematics teachers, we have become increasingly aware of the pervasive influence teachers' images have on how they implement innovative curricula. We have observed that these images manifest themselves in two sharply contrasting orientations toward mathematics teaching. We refer to these orientations as *calculational* and *conceptual.* To illustrate what we mean by a calculational and a conceptual orientation in teaching mathematics, we start with two vignettes. After the vignettes we discuss more generally what these orientations entail and their implications for classroom discourse and for students' learning. The article ends with a discussion of obstacles to adopting a conceptual orientation and a discussion of the implications these obstacles have for the professional preparation and development of mathematics teachers.

The vignettes depict two different teachers, each illustrative of an orientation. Our intent is to give the reader concrete examples of the kind of teaching—specifically the nature of the classroom discourse—that is characteristic of each orientation. The vignettes have been constructed from videotaped observations of actual lessons.

Research reported in this paper was supported by National Science Foundation (NSF) Grants Nos. MDR 89–50311 and 90–96275 and by a grant of equipment from Apple Computer, Inc., Office of External Research. Any conclusions or recommendations stated here are those of the authors and do not necessarily reflect official positions of NSF or Apple Computer.

Vignette 1

A seventh-grade teacher presents the following problem to his class:

> At some time in the future John will be 38 years old. At that time he will be 3 times as old as Sally. Sally is now 7 years old. How old is John now?

After allowing students time to think about the problem and to discuss their thinking with a classmate, the teacher calls for volunteers to explain how they thought about the problem in order to solve it. What follows are the responses offered by the students and the ensuing exchange between teacher and students:

T: Let's talk about this problem a bit. How is it that you thought about it?

S1: I divided 38 by 3 and I got 12 2/3. Then I subtracted 7 from 12 2/3 and got 5 2/3. (*Pause*) Then I subtracted that from 38 and got 32 1/3. (*Pause*) John is 32 1/3.

T: That's good! (*Pause*) Can you explain what you did in more detail? Why did you divide 38 by 3?

S1: (*Appearing puzzled by the question, S1 looks back at her work. She looks again at the original problem.*) Because I knew that John is older—3 times older.

T: Okay, and then what did you do?

S1: Then I subtracted 7 and got 5 2/3. (*Pause*) I took that away from 38, and that gave me 32 1/3.

T: Why did you take 5 2/3 away from 38?

S1: (*Pause*) To find out how old John is.

T: Okay, and you got 32 1/3 for John's age. That's good! (*Pause*) Yes, S2?

S2: Isn't the answer 21? (*Pause*) I multiplied 7 times 3 and I got 21.

T: Hum? Not quite. (*Pause*) How come you multiplied 7 times 3?

S2: It says that he is 3 times as old as Sally ... (*Pause*) and Sally is 7.

T: Oh, I see! (*Pause*) You're right, the problem says that John is 3 times as old as Sally, but that is when John is 38. That's at the time he is 38, which is at some time in the future. (*Pause*) Do you understand?

S2: Sort of.

T: Okay, how about you, S3? How did you think about it?

S3: I divided 38 by 3, and I subtracted that from 38. That's 25 and something. Then I added that to 7. I got the same thing as S1—32 something.

T: But you did it differently. Super! See? There are different ways to solve the same problem. (*Pause*) How about you, S4?

S4: I subtracted 7 from 38 and divided that by 3. (*Pause*) I got 10 something. Then I added that to 7. (*Pause*) I got that he is 17 and something.

T: Hum? That doesn't quite agree with the other answers, does it? I'm not sure I understand what you're doing. (*Pause*) Why did you subtract 7 from 38?

S4: (*Shrugging his shoulders*) I don't know.

T: S5?

S5: Dividing 38 by 3 can't be right! It doesn't come out even.

T: That doesn't matter, does it? We still get a number, don't we? (*Pause*) We get that Sally is 12 2/3. (*Pause*) Let's take a look at how to divide 38 by 3. Divide 3 into 38. (*Motioning with his hands in the air as if he were doing the long division on an imaginary chalkboard*) Three goes into 38 ten times, put up the 1, and 10 times 3 is 30. Thirty-eight minus 30 is 8. Three goes into 8 two times. Put up the 2, and 2 times 3 is 6. So 8 minus 6 is 2. The answer is 12 remainder 2, or 12 and 2/3. Okay? (*Pause*) Let's take a look at the two ways the problem was solved.

The teacher proceeds to demonstrate S1's and S3's solutions on the chalkboard and refers to both solutions as appropriate ways to think about the problem. This segment of the lesson ends and the class moves to work on another task.

Contrast the vignette given above with the one below, which illustrates an exchange of a very different nature between a teacher and his students. This exchange followed the presentation of the same problem as in vignette 1 to a group of seventh graders. Again, the exchange takes place after the students have had the opportunity to think about the problem and to discuss it with a classmate.

Vignette 2

T: Let's talk about this problem a bit. How is it that you thought about the information in it?

S1: Well, you gotta start by dividing 38 by 3. Then you take away ...

T: (*Interrupting*) Wait! Before going on to tell us about the calculations you did, explain to us why you did what you did. (*Pause*) What were you trying to find?

S1: Well, you know that John is 3 times as old as Sally, so you divide 38 by 3 to find out how old Sally is.

T: Do you all agree with S1's thinking?

Several students say, "Yes"; others nod their heads.

S2: That's not gonna tell you how old Sally is *now*. It'll tell you how old Sally is when John is 38.

T: Is that what you had in mind, S1?

S1: Yes.

T: (*To the rest of the class*) What does the 38 stand for?

S2: John's age in the future.

T: So 38 is not how old John is now. It's how old John will be in the future. (*Pause*) The problem says that when John gets to be 38 he will be 3 times as old as Sally. Does that mean "3 times as old as Sally is *now*" or "3 times as old as Sally will be when John is 38"?

Several students respond in unison, "When John is 38."

T: Are we all clear on S2's reasoning? (*Pause*)

S3: I started the same way, but I got stuck dividing. (*Pause*) Three doesn't go into 38 evenly. (*Pause*)

T: Don't worry about how to divide 38 by 3 now. That's not what's most important right now. What are you trying to find by dividing 38 by 3?

S3: Sally's age.

T: Sally's age when John is 38 years old. (*Pause*) You can use your calculator if you want to. (*Pause*) If you try it, you'll get 12.66 … years. That's Sally's age in the future. (*Pause*) S4?

S4: Couldn't you just say John is 21? (*Pause*) Couldn't you just multiply 3 times 7?

T: What will that give you?

S4: Twenty-one!

T: Yes, I know that. But what would the 21 represent? What is it that's 21?

S4: That's how old John is now. Isn't that what we want to find?

S5: No! (*Pause*) I mean, yes! That's what we want to find, but that's not right!

T: What is it that is not right, S4? We do want to find out how old John is now, don't we?

S5: Right. But see, he's not 3 times older than Sally *now!* He'll be 3 times older than Sally when he is 38. So you can't multiply 7 by 3.

T: Let's think about that. If we know that John will be 3 times as old as Sally when he is 38, does that make him 3 times as old as Sally now? (*Pause*) S4, what do you think?

S4: I guess not. (*Pause*)

T: (*To S4*) Suppose you're now 12 and your younger sister or brother is 6 years old. That makes you twice as old as your younger sister. Will that also be true next year? (*Pause*) Next year you'll be 13 and she'll be 7. Will you still be twice as old as your sister?

S6: Actually, that'll happen only once and never again.

S4: I see it.

T: Okay, so how are we going to use the information that John will be 3 times as old as Sally when he gets to be 38? (*Pause*) Who can explain?

S1: You can divide 38 by 3 and get 12.66….

T: Remember to tell us what your numbers stand for. What does the 12.66 … stand for?

S2: That's how old Sally will be.

T: When?

Several students respond, "When John is 38."

T: Okay, we know how old Sally will be when John is 38 years old. (*Pause*) She will be 12.66 … years. We can say she'll be 12, because we usually don't say that we are 12.66 … years old. We typically use whole numbers when we talk about our age. Okay?

S6: Okay, you can say that Sally will be 12. So if you subtract 7 from that you get 5. Then you take away 5 from 38 and you're done! John is 33.

T: Wait a minute! You're going too fast. I don't see how you know to do all that. Can you explain your reasoning?

S6: (*Patiently*) You know Sally will be 12 and something, and you know that she is 7 now. So that means that there are 5 years between now and then. Actually a little more than 5 years, but you said that was okay.

T: Yes, it's okay to say 5 years. So 5 years is how much time there is between now and the time in the future when John is 38?

S6: Yes, So if you take 5 away from 38, that's how old John is now.

T: Did everyone follow S6's reasoning? (*Pause*) Who can recap the solution we've just been through?

The teacher calls on two volunteers who, with some assistance from other classmates and the teacher, summarize the discussion.

T: Did anyone think about the problem differently? (*Pause*) S7?

S7: Well, sort of. I started out the same. I divided 38 by 3.

T: (*Interrupting*) To find what?

S7: Sally's age in the future.

T: Okay.

S7: I got that Sally will be 12 2/3 years old when John is 38. Then I subtracted to find the difference between their ages. (*Pause*) I got 25 1/3.

T: Twenty-five and one-third what?

S7: Twenty-five and one-third years. That's how much older John is. (*Pause*)

T: How much older than Sally?

S7: Yes. That's the difference between their ages.

T: Now or when John is 38?

S7: Actually, it doesn't matter. The difference between their ages will always be the same.

T: Okay, we can come back to that thought in a minute. (*Pause*) Go on.

S7: So to find out how old John is now…. See, you know Sally is now 7 and John is 25 1/3 years older than Sally. So add 25 1/3 to 7, and you get John's age. That's 32 1/3. (*Pause*) That's how I figured it.

T: Who agrees with S7's reasoning?

Several hands go up.

S8: I don't understand why she added 25 1/3.

S2: Because that's how much older John is than Sally.

S8: I still don't see why she added that to 7.

S2: If you know Sally is 7 and John is 25 1/3 years older than Sally, you add to get how old John is now.

S8: (*Puzzled*) But 25 1/3 is when John is 38 and Sally is 12 2/3.

S9: The difference between their ages is always the same—now and when John is 38.

T: Does that make sense to everyone? (*Pause*) Who can explain S7's solution method from the beginning? (*Pause*) Don't just tell me what operations she did. Remember, "to explain" means that you have to talk about her reasoning, not just the arithmetic she did.

The discussion continues. The teacher poses more questions aimed at focusing students' attention on the quantities and quantitative relationships in the problem. He probes for the reasoning underlying the students' arithmetic procedures. As the teacher elicits responses from the students, he sketches a diagram (fig. 8.1) to support the discussion of invariance of age differences and variance of age ratios.

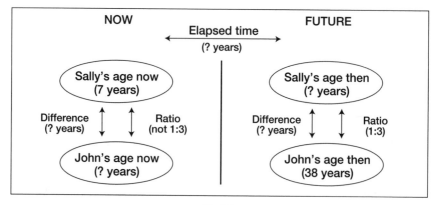

Fig. 8.1

THE VIGNETTES AND THE TEACHERS: SIMILARITIES AND DIFFERENCES

We constructed the vignettes from actual classroom observations to capture as concretely as possible what we have observed to be important differences in mathematics classroom discourse. Despite the obvious similarities, important substantive differences can be discerned in the two vignettes. Although both teachers opened their lessons with the same problem and with similar instructions, the ensuing discussions were quite different. They not only differed in superficial, albeit important, features, such as duration and number of students involved (the discussion of the problem in the first vignette was much briefer than in the second vignette, and vignette 1 overtly involved five students, whereas nine students contributed to the discussion in vignette 2), but also differed markedly in what was discussed and in the roles the teachers played.

In both vignettes the students initially offered sequences of arithmetic procedures as expressions of their thinking. However, in vignette 2 the students began to give explanations that were grounded in conceptions of the situation. In contrast, the explanations given by the students in vignette 1 remained strictly procedural; they were all statements of how they calculated John's age, and they all failed to address what the teacher ostensibly requested—an explanation of how they thought about the problem. The students in vignette 1 did not offer a justification for the chosen operations that was grounded in conceptions of the situation; when explaining, they did not connect their calculations to the ideas of time, duration, aging, or relationships among them. Their explanations were "calculational," which stand in sharp contrast to the conceptual explanations given in vignette 2.

Both teachers pressed their students to give rationales for their calculational solutions, but they did so differently and with quite different results. When compared with the explanations elicited by teacher 2, the explanations obtained by teacher 1 were shallow and incomplete (recall the student who justified dividing 38 by 3 by saying that John is older than Sally). Teacher 1 was less persistent than teacher 2 in probing the students' thinking. He accepted solutions consisting of calculational sequences if they were correct by some criteria that he did not make explicit to the students. Teacher 2, in contrast, persistently probed students' thinking whenever their responses were cast in terms of numbers and operations, thus steering the discussion and focusing the students' attention on how they were conceiving the situation. His students were more inclined to comment on one another's contributions than were teacher 1's students.

Another important difference between the teachers was in their responses to the students' difficulties with dividing 38 by 3. Teacher 1 used the occasion as an opportunity to review the long-division algorithm; teacher 2 steered the students' attention away from the computational difficulty, downplaying its significance and redirecting their attention toward the quantitative relationship that suggested division.

The actions of the two teachers were driven by different images of their pedagogical tasks and of the goals they served. Teacher 1's image was that there was a problem to be solved. Teacher 2's image was of an occasion for students to reason and to reflect on their reasoning. Although it might be argued that for both teachers the general goal was the long-term development of the students' problem-solving skills, for teacher 2 that development clearly entailed getting the students skilled at reasoning. Furthermore, teacher 2 had an image of what is involved in becoming a skilled reasoner, which he obviously had translated into specific pedagogical practices. His actions appeared to be driven by the belief that not until students make their reasoning explicit to themselves can they reflect on it and represent it mathematically and that those representations empower their reasoning. The distinctions between these teachers' actions reside in their orientations toward mathematics and teaching mathematics. The teacher in vignette 1 exemplifies what we call a "calculational orientation." The teacher in vignette 2 exemplifies what we call a "conceptual orientation."

In the remainder of this paper we focus on these two orientations from a more theoretical perspective. First, we characterize the two orientations. Next, we address the consequences of each orientation for the teachers' instructional practices, the students' learning and beliefs, and the nature of the classroom discourse. We conclude with a discussion of the obstacles to adopting a conceptual orientation and some remarks about what might be involved in successfully adopting such an orientation.

TWO CONTRASTING ORIENTATIONS

We believe that the substantive differences in the way the teachers handled the curricular task in the vignettes are an expression of a fundamental difference in their orientations toward teaching mathematics. As mentioned above, we refer to these as "conceptual" and "calculational" orientations. Here is how we characterize them.

A teacher with a conceptual orientation is one whose actions are driven by—

- an image of a system of ideas and ways of thinking that she intends the students to develop;
- an image of how these ideas and ways of thinking can develop;
- ideas about features of materials, activities, and expositions and the students' engagement with them that can orient the students' attention in productive ways (a productive way of thinking generates a "method" that generalizes to other situations;
- an expectation and insistence that students be intellectually engaged in tasks and activities.

Conceptually oriented teachers often express the images described above in ways that focus students' attention away from the thoughtless application of procedures and toward a rich conception of situations, ideas, and relationships among ideas. These teachers strive for conceptual coherence, both in their pedagogical actions and in students' conceptions. As a result, conceptually oriented teachers tend to focus on aspects of situations that, when well understood, give meaning to numerical values and that suggest numerical operations (Thompson in press). Conceptually oriented teachers often ask questions that move students to view their arithmetic in a noncalculational context like the following:

- "(This number) is a number of what?"
- "To what does (this number) refer in the situation we're dealing with?"
- "What are you trying to find when you do this calculation (in the situation as you currently understand it)?"
- "What did this calculation give you (in regard to the situation as you currently understand it)?"

The actions of a teacher with a calculational orientation are driven by a fundamental image of mathematics as the application of calculations and procedures for deriving numerical results. This does not mean that such a teacher is focused only on computational procedures.[1] Rather, his view of mathematics is more inclusive but still one focused on procedures—computational or otherwise—for "getting answers."

These are some symptoms of a calculational orientation:

- A tendency to speak exclusively in the language of numbers and numerical operations
- A predisposition to cast solving a problem as producing a numerical solution

---

1. This view we call a "computational orientation." A teacher with a computational orientation views mathematics as composed of computational procedures, and doing mathematics as computing in the absence of any reason for the computation aside from the context of having been asked to do so. A computational orientation implies a calculational orientation but a calculational orientation does not imply a computational orientation.

- An emphasis on identifying and performing procedures
- A tendency to do calculations whenever an occasion to calculate presents itself regardless of the overall context in which the occasion occurs
- A tendency to disregard the context in which the calculations might occur and how they might arise naturally from an understanding of the situation itself
- An inclination to remediate students' difficulties with calculational procedures independently of the context in which the difficulties manifest themselves
- A tendency to treat problem solving as flat; that is, nothing about problem solving is any more or less important than anything else, except that the answer is most important because getting the answer is the reason for solving the problem
- A narrow view of mathematical patterns as limited to finding patterns in numerical sequences and in the sameness of operations across problems, as opposed to finding patterns in reasoning in the solution of problems

Consequences of Calculational and Conceptual Orientations

Calculational and conceptual orientations can have different consequences for the interchanges that occur in classrooms (Wertsch and Toma in press). These consequences can be organized by the interplay between teachers and students according to which orientations each possesses and by the interplay among students possessing different orientations. We shall focus on the influence of teachers' orientations on classroom discourse because we believe that teachers set the tone for the kinds of discussions in which students engage, whether with the teacher or among themselves (Cohen 1990; Porter 1989; A. Thompson and P. Thompson 1994; P. Thompson and A. Thompson in press).

The teachers' goals and images described in the previous section account for many of the differences between the two vignettes. The first teacher's goal was for students to solve the problem and share their procedures; the second teacher's goal was to create an occasion for students to reason and to make their reasoning public. Subtle but important differences in the teachers' behaviors were an expression of their different goals.

In the previous section we described the teachers' pedagogical tasks. The teacher in vignette 1 expected his students to explain their procedures; the teacher in vignette 2 expected students to explain their reasoning. One manifestation of the teachers' goals is the type of questions they asked. For example, both teachers asked S1 why she had decided to divide 38 by 3. The second teacher also asked S1, "What were you trying to find when you divided 38 by 3?" By asking this question, the teacher oriented his students toward the situation itself and their conception of it, which required the students to reflect on their understanding of the situation. An important feature of vignette 2 is that the teacher persisted in bringing students back to thinking about their conceptions of the situation. This orientation contrasts with orienting students to reflect on their calculations and with allowing students to remain oriented toward their calculations.

Students also have varying degrees of conceptual or calculational orientations to mathematics. Those who have adapted to calculationally oriented instruction will approach mathematical discussions with the expectation that they will be concerned with getting answers (Cobb, Yackel, and Wood 1989; Nicholls et al. 1990). Students who have come to view mathematics as "answer getting" not only will have difficulty focusing on their and others' reasoning but also may consider such a focus as being irrelevant to their images of what mathematics is about.

Conversely, students who have adopted a conceptual orientation will likely engage in longer, more meaningful discussions (Cobb, Wood, and Yackel 1991). Vignette 2 lasted longer and involved more students than vignette 1 because students had something to discuss. Students in vignette 1 did not sustain a substantive discussion because they had no way of knowing the sources of their classmates' procedures. Reasoning was not a subject to discuss. Students in vignette 2, through the support of their teacher, did discuss their reasoning and in doing so created an environment in which they felt free to share their understandings.

A calculationally oriented teacher may believe that explaining the calculations one has performed is tantamount to explaining one's reasoning (Cobb, Wood, and Yackel in press). We observed that the only students able to follow a calculational explanation are those who understood the problem in the first place and understood it in such a way that the proposed sequence of operations fits their conceptualization of the problem. To illustrate this observation, imagine four students, Alicia, Betty, Carl, and Don, all of whom solved the "Sally and John" problem incorrectly. Furthermore, imagine that their errors stemmed from different sources. Alicia missed the problem because she committed a calculational error, but her understanding of the problem was valid and she understood the problem in a way that fit the calculational explanation offered by S1. Betty missed the problem because of a calculational error; her understanding of the problem was valid, but her understanding of the problem did not fit the string of calculations offered by S1. Carl and Don missed the problem because they could not conceptualize it; Don possesses a calculational orientation and Carl possesses a conceptual orientation. The four students are listening to the discussion between S1 and T1 (the teacher in the first vignette):

S1: I divided 38 by 3 and I got 12 2/3. Then I subtracted 7 from 12 2/3 and got 5 2/3. Then I subtracted that from 38 and got 32 1/3. John is 32 1/3.

T1: That's good! (*Pause*) Can you explain what you did in more detail? Why did you divide 38 by 3?

S1: (*Appearing puzzled by the question, she looks back at her work. She looks again at the original problem.*) Because I knew that John is older—three times older.

T1: Okay, and then what did you do?

S1: Then I subtracted 7 and got 5 2/3. (*Pause*) I took that away from 38 and that gave me 32 1/3.

T1: Why did you take 5 2/3 away from 38?

S1: (*Pause*) To find out how old John is.

T1: Okay, and you got 32 1/3 for John's age. That's good!

For Alicia, who had made a calculational error but understood the problem in a way that fits S1's string of operations, this explanation validates her solution attempt, leaving her with the sense that she now understands what she had actually understood all along. Betty is convinced that she does not understand the problem at all—her initial answer was incorrect, and S1's string of operations do not fit with the way she conceived the problem. Don thinks he now understands, since he was able to follow all S1's calculations. Don's ability to perform all the calculations may even give him the confidence to explain S1's solution to Carl, who complains that he does not understand. However, Don's procedural explanation only leaves Carl even more frustrated, since he finds Don's explanation incomprehensible. These explanations do not tell Carl why the calculations were performed. In fact, with all the Dons in the class nodding as if they understand, Carl may judge that there is something wrong with his ability to understand mathematics, when in fact the only problem is that his expectations for understanding are greater than those of his peers. Over time, a conceptually oriented student such as Carl, sitting in a classroom dominated by calculationally oriented discourse, may conclude that mathematics is not supposed to make sense. Eventually, he may altogether stop trying to understand mathematics.

OBSTACLES AND IMPLICATIONS

It is evident to us that a conceptual orientation is by far the more enriching and the more productive for students and for teachers. But this orientation cannot be created easily, nor once created, can it be easily maintained (Romberg and Price 1981; von Glasersfeld 1988; Wood, Cobb, and Yackel 1991). To create a conceptual orientation, the teacher must reflect long and deeply on her goals for, and images of, mathematics and mathematics teaching. In our personal experience, there are periods of confusion about what we are trying to have our students understand, and teachers working with us have expressed the same feelings. When we move our focus of instruction to deep conceptualizations of situations, we also move away from the domains of discourse with which we feel most comfortable—established methods for deriving numerical solutions. Instead, we move toward domains of discourse that emphasize "how you think about it"—domains few of us have explored and too few students have experienced.

One of the major obstacles to creating a conceptual orientation is teachers' lack of ideas about how to move pedagogically from holding conversations about "how you think about it" to the standard mathematics of the conventional curriculum. Teachers frequently ask us essentially this question: "After we've talked about understanding these situations, how do I introduce the standard procedures?" This question indicates to us a teacher who is grappling with a dilemma—how to reconcile an emphasis on students' reasoning with the traditional curriculum and pedagogy wherein symbols, methods, and procedures are introduced before students encounter any substantive applications.

A conceptual approach to teaching mathematics aims for students to solve problems by working from a deep understanding of them. But working from an understanding means that they work from *their* understandings. A primary aim of conceptually oriented teaching is that students come to conceive a conceptual domain by developing methods for solving problems in it. Part of students' developing stable, general methods is that they deal with the matter of expressing those methods in notation. Once students have developed conceptual methods and have reflected those methods in notation, they can then appreciate that conventional methods are but one way to solve problems in a conceptual domain. It is important that students also appreciate that the most powerful approach to solving problems is to understand them deeply and proceed from the basis of understanding and that a weak approach is to search one's memory for the "right" procedure. A teacher's dilemma regarding when to introduce conventional procedures is eventually resolved when this teacher realizes that no reconciliation is possible; the traditional curriculum turns the construction of mathematical meaning upside down. The resolution of the dilemma comes from the teacher's creation of a new philosophy—a philosophy of what he or she is trying to attain that permeates his or her instructional goals and actions (Ball 1993).

Once a teacher makes a commitment to treat mathematics conceptually, she loses the support structures on which she has come to rely, such as textbooks and repertories of stable practices. This loss is very threatening and thus is a major obstacle to change. Old habits die hard and new practices evolve slowly. For most teachers who lack the time and energy to rethink their curriculum and pedagogy, the thought of giving up conventional materials can be very unsettling. Our research suggests that having a repository of rich problems is enough to begin moving away from the textbook. Our research also suggests that such a repository is not sufficient to ensure success; a conceptual understanding of the subject matter the problems address is also necessary for teachers to feel that they have a sense of direction and to be able to respond to students' difficulties.

To teach mathematics conceptually, it is not sufficient to know how to solve the problem with which the students may be grappling, nor even to know several methods of solution (McDiarmid, Ball, and Anderson 1989). To teach conceptually requires a deep conceptualization of the situation. This, in turn, requires that the teacher think beyond what is necessary to simply find ways of dealing mathematically with the situation. Furthermore, to be able to orient students' thinking in productive ways, it is extremely helpful to have an image of students' thinking as they develop these ideas. Any teacher can begin building this image by encouraging students to reason and express themselves accordingly and by listening to their reasoning, respecting it, and asking the other students to do likewise.

REFERENCES

Ball, Deborah L. "With an Eye on the Mathematical Horizon: Dilemmas of Teaching Elementary School Mathematics." *Elementary School Journal* 93 (March 1993): 373–97.

Bauersfeld, Heinrich. "Hidden Dimensions in the So-Called Reality of a Mathematics Classroom." *Educational Studies in Mathematics* 11 (February 1980): 23–42.

Cobb, Paul, Terry Wood, and Erna Yackel. "Classrooms as Learning Environments for Teachers and Researchers." In *Constructivist Views on the Teaching and Learning of Mathematics* (*Journal for Research in Mathematics Education* Monograph No. 4), edited by Robert B. Davis, Carolyn A. Maher, and Nel Noddings, pp. 125–46. Reston, Va.: National Council of Teachers of Mathematics, 1991.

———. "Discourse, Mathematical Thinking, and Classroom Practice." In *Contexts for Learning: Social Cultural Dynamics in Children's Development,* edited by E. Forman, N. Minick, and A. Stone. Oxford: Oxford University Press, in press.

Cobb, Paul, Erna Yackel, and Terry Wood. "Young Children's Emotional Acts While Engaged in Mathematical Problem Solving." In *Affect and Mathematical Problem Solving,* edited by D. B. McLeod and V. A. Adams, pp. 117–48. New York: Springer-Verlag, 1989.

Cohen, David K. "A Revolution in One Classroom: The Case of Mrs. Oublier." *Educational Evaluation and Policy Analysis* 12 (Fall 1990): 327–45.

Cooney, Thomas J. "A Beginning Teacher's View of Problem Solving." *Journal for Research in Mathematics Education* 16 (November 1985): 324–36.

McDiarmid, G. W., Deborah L. Ball, and C. W. Anderson. "Why Staying One Chapter Ahead Doesn't Really Work: Subject-Specific Pedagogy." In *Knowledge Base for Beginning Teachers,* edited by M. C. Clinton, pp. 193–206. New York: Pergamon Press, 1989.

Nicholls, John, Paul Cobb, Erna Yackel, Terry Wood, and Grayson Wheatley. "Students' Theories about Mathematics and Their Mathematical Knowledge: Multiple Dimensions of Assessment." In *Assessing Higher Order Thinking in Mathematics,* edited by Gerald Kulm, pp. 137–54. Washington, D.C.: American Association for the Advancement of Science, 1990.

Porter, Andrew. "A Curriculum Out of Balance: The Case of Elementary Mathematics." *Educational Researcher* 18 (June-July 1989): 9–15.

Romberg, Thomas A., and G. Price. "Assimilation of Innovations into the Culture of Schools: Impediments to Radical Change." Paper presented at the National Institute of Education, Washington, D.C., May 1981.

Thompson, Alba G. "The Relationship of Teachers' Conceptions of Mathematics Teaching to Instructional Practice." *Educational Studies in Mathematics* 15 (May 1984): 105–27.

Thompson, Alba G., and Patrick W. Thompson. "Talking about Rates Conceptually, Part I: A Teacher's Struggle." *Journal for Research in Mathematics Education* 25 (May 1994): 279–303.

Thompson, Patrick W., and Alba G. Thompson. "Talking about Rates Conceptually, Part II: Pedagogical Content Knowledge." *Journal for Research in Mathematics Education,* in press.

Thompson, Patrick W. "Quantitative Reasoning, Complexity, and Additive Structures." *Educational Studies in Mathematics,* in press.

von Glasersfeld, Ernst. "Reluctance to Change a Way of Thinking." *Irish Journal of Psychology* 8 (1)(1988): 83–90.

Wertsch, James V., and Chikako Toma. "Discourse and Learning in the Classroom: A Sociocultural Approach." In *Constructivism in Education,* edited by Leslie P. Steffe and Jerry Gale, Hillsdale, N.J.: Lawrence Erlbaum Associates, in press.

Wood, Terry, Paul Cobb, and Erna Yackel. "Change in Teaching Mathematics: A Case Study." *American Educational Research Journal* 28 (Fall 1991): 587–616.

9

Teachers as Assessors: A Professional Development Challenge

Ralph W. Cain
Patricia Ann Kenney
Cathy G. Schloemer

THE NCTM's *Curriculum and Evaluation Standards for School Mathematics* (1989) and *Professional Standards for Teaching Mathematics* (1991) together present a vision for the teaching and learning of mathematics. Much of this vision is presented in terms of *change*—change from an emphasis on rote memorization of facts and procedures to an emphasis on problem solving; from teachers as dispensers of information to teachers as facilitators of student learning; and from dependence on paper-and-pencil tests to the incorporation of alternative assessment methods. It is important to recognize that much of the responsibility for implementing change falls squarely on the shoulders of classroom teachers as "key figures in changing the ways in which mathematics is taught and learned in schools" (NCTM 1991, p. 2). To bolster the assumption that "such changes require that teachers have long-term support and adequate resources" (p. 2), the *Professional Teaching Standards* includes a section on professional development that suggests ways to provide mathematics teachers at all experience levels with the knowledge and skills necessary to develop, maintain, and improve the quality of their teaching.

One of the specific areas that both the *Curriculum and Evaluation Standards* and the *Professional Teaching Standards* have targeted for change is classroom assessment. These documents propose a vision that requires teachers to use assessment methods that go far beyond paper-and-pencil tests and traditional homework assignments. This movement toward alternative, authentic classroom assessment has been advocated not only by NCTM and individual members of the mathematics education community (e.g., Clarke, Clarke, and Lovitt 1990; Lester and Kroll 1991; Stenmark 1991; Webb 1993; Webb and Briars 1990) but also by educators from many other subject fields (e.g., Archbald and Newmann 1988; Baron 1990; Linn 1990; Resnick and Resnick 1992; Shepard 1989). The classroom teacher of mathematics, then, must become an assessor, a person responsible for the *development* of appropriate assessment methods, the *interpretation* of assessment information, and the *communication* of assessment results (Cain and Kenney 1992).

Ensuring that teachers can function effectively as assessors is an important professional development challenge to the mathematics education community. The purpose of this article is to explore the assessment responsibilities that teachers of mathematics have and to suggest formal and informal professional development activities that will help them become successful assessors. Although there are a variety of ways in which teachers can engage in professional development, this article focuses on activities that emphasize teachers sharing assessment expertise and ideas with their fellow teachers.

THE DEVELOPMENT OF ASSESSMENT METHODS

The notion that teachers are also "assessment developers" is not really new. Each day teachers must create a variety of ways to gather information about their students' mathematical knowledge and processes. For example, because tests and other assessment instruments available from external sources (e.g., testing companies, textbook publishers) may not appropriately address teachers' needs or those of their students, teachers often design quizzes and unit tests to supplement—or even replace—ones that come "ready made." Even more common than written tests are the observation and questioning opportunities that routinely appear as part of normal classroom discourse.

The teacher's role as assessment developer assumes additional responsibilities in the mathematics classroom in which curriculum and instruction are aligned with the goals of the NCTM *Standards* documents. Measuring students' growth in mathematical power requires assessment to be "a continuous, dynamic, and often informal process, ... a 'biography' of student learning" (NCTM 1989, p. 203). Teacher-produced written tests and informal observation and questioning techniques are likely to contribute important information to this biography; however, open-ended tasks, extended investigations, portfolios, student self-assessment activities, and journals are additional assessment methods that have been suggested as appropriate ways to measure mathematical power. Thus, teachers of mathematics at all grade levels must be able to develop appropriate alternative assessments and to ensure that the assessments serve their intended purposes.

The role of assessment developer raises a number of important issues, one of which involves the teacher's familiarity with alternative methods. If preservice teachers had been required to take a "tests and measurements" course, they most likely took one in which paper-and-pencil and multiple-choice tests were emphasized to the exclusion of nontest, alternative assessment methods (Goehring 1973; Gullickson 1986). If it is true, then, that teachers of mathematics have had little or no formal training in developing alternative assessments, then the challenge becomes one of creating opportunities for them to learn to do so.

One way that teachers can enhance their skills in assessment development is to become aware of the abundance of information on mathematics assessment that is available. NCTM publications (e.g., Charles, Lester, and O'Daffer 1987; Elliott 1992; Stenmark 1991; Webb 1993), reports of state assessment programs (e.g., California State Department of Education 1989; Mumme 1990; Pandey 1991; Vermont Department of Education 1991), and materials from other countries (e.g., Clark 1992; Clarke 1988; Shell Centre for Mathematical Education 1984) showcase the use of alternative assessment in the mathematics classroom and provide a rich variety of examples. Teachers can use these sources to get ideas for alternative assessments (e.g., checklists, questioning techniques, open-ended tasks, extended investigations, portfolios, journals) and then can adapt them to conform to their own needs and those of their students.

The scenario presented above is predicated on the assumptions that teachers have somehow personalized the need for their own professional development in the area of mathematics assessment and that they know where to find assessment ideas. But what can be done for teachers who need assistance in thinking about ways to create effective alternative assessments? Clearly, some form of external intervention is necessary, but the intervention is likely to be most effective if it comes from a local source. For example, an administrator, department head, or teacher who has been reading the literature on alternative mathematics assessment might share an article with other teachers on a regular basis. Teachers who have attended conference presentations or who have taken courses on alternative assessment could find ways to share their new-found knowledge. Or perhaps a well-respected teacher who has tried to design a portfolio assessment program could be recruited to serve as a mentor to fellow teachers. In fact, it may be appropriate for these mentor teachers to come from other curricular areas. English and language arts teachers have had much experience in developing holistic writing assessments, and their ideas and experiences could be shared with mathematics teachers, who most likely have had little or no experience in developing these types of activities. This idea of fostering professional development through a "sharing" experience among teachers, supervisors, and administrators can encourage serious thinking about the development of alternative mathematics assessment methods.

THE INTERPRETATION OF ASSESSMENT INFORMATION

Just as developing assessment methods can be thought of as a common activity for mathematics teachers, so interpreting assessment information is something that teachers do almost every day. Throughout the school day, teachers can instantaneously interpret what they learn from observing and listening to their students, or they may choose to reflect on this information during a planning period. For more formal assessments, mathematics teachers have additional interpretive responsibilities that include establishing scoring criteria and translating the results into a letter or numerical grade.

The interpretation of assessment information is dependent on its use. Both the NCTM *Curriculum and Evaluation Standards* and the *Professional Teaching Standards* advocate that the primary use for assessment information is instructional guidance, the "shaping and reshaping of the learning environment of the classroom" (NCTM 1991, p. 63). Although this is a worthy position to take, a much more concrete concern involves the translation of the information from alternative assessment methods into a form that is appropriate for providing feedback to students, particularly in the form of a quantified score that is useful for grading purposes. For the more traditional assessment methods, such as quizzes, homework, and unit tests, basing the measure of performance on the percent of problems correctly answered or on a point-based, partial-credit scheme has been generally accepted by teachers and understood by their students and other interested parties (e.g., parents, guardians, school administrators). However, with alternative assessment methods, quantitative or qualitative interpretation of student work may be foreign territory to most teachers. For example, it would be inappropriate and even difficult to assign a "score" to an interview, a portfolio, a journal, or a cooperative group project using percent-correct or partial-credit methods. Instead, results from alternative assessments are best obtained through the use of holistic scoring rubrics and checklists. Issues of subjectivity, reliability, and fairness, as well as the determination of individual contributions to group activities and projects, and the translation of these into numerical scores for purposes of assigning grades are possible sources of conflict and, in extreme cases, legal actions. It becomes apparent that teachers, if they are to implement alternative assessment methods in their classrooms, need to be skilled in interpreting the information derived from alternative assessments so that it can be used to provide meaningful instructional guidance and student feedback as well as to assign fair and valid grades.

Teachers who have become familiar with the publications and reports mentioned previously will find that they also contain examples of ways to interpret information from alternative assessments. One of the most common of these examples involves the use of scoring guides to transform student performance on open-ended mathematics tasks into a numerical score. The inclusion of sample student responses that are representative of performance at each scoring level can assist teachers in understanding how to use a scoring guide. Both practicing and preservice teachers could benefit from professional development activities that present information on holistic evaluation methods and the use of scoring guides and student responses. A suggested activity draws on the expertise of English and language arts teachers, who, in addition to furnishing information on the development of writing assessments for mathematics, could also share their expertise on grading students' writing holistically. With respect to the use of scoring guides, a group of mathematics teachers could discuss prescored student responses in relation to the scoring guide, thus internalizing the criteria that were used in evaluating students' work. Another form of this activity that takes on a "constructivist" flavor would be for a group of teachers to write one or two open-ended mathematics tasks, administer them to their students, and then evaluate the responses using an agreed-on scoring scheme.

As teachers become aware of the various interpretive methods for alternative assessments, it is important that they extend their professional development to include "trying out" some specific methods in their mathematics classrooms. Suggestions for ways to collect and interpret this information are beginning to appear in articles and other publications. For example, Kroll, Masingila, and Mau (1992) have suggested ways in which information gathered from cooperative group work can be transformed into grades. Perhaps a teacher who has used this article as a basis for grading group work can share the strengths and limitations of this system either through informal discussions in the teachers' lounge or through a formal in-service presentation. As teachers become more experienced and well versed in the interpretation of information from alternative assessments, they have the potential to become excellent on-site resource persons to their fellow teachers.

THE COMMUNICATION OF ASSESSMENT RESULTS

An important part of any assessment involves communicating the results to interested parties. Mathematics teachers have the responsibility of sharing this information with students, parents or guardians, and school administrators through a variety of methods such as giving report card grades, communicating with parents or guardians through notes or telephone calls, and discussing results with supervisors and curriculum coordinators. It is important that all sharing be done with regard to the appropriate and ethical use of assessment results. For example, teachers should consider the needs and feelings of students before releasing results to, or sharing them with, anyone, and results should be released only to those persons who truly have a need to know. An additional communication responsibility arises when a teacher's evaluation method is challenged by students, parents, administrators, or in extreme situations through the legal system.

With respect to sharing results from alternative assessments, mathematics teachers may find themselves undertaking a different, somewhat difficult task. The instructional program based on the NCTM *Standards* documents is quite different from earlier programs. As a result, it is almost certain that grading will be based on results from alternative assessment methods as well as those from more traditional methods. Parents, guardians, and certainly students may not be familiar with techniques that a teacher used to transform information from observations, journals, group work—as well as from quizzes and unit tests—into a single numerical grade (e.g., "82") or grade symbol (e.g., "B"). When one considers the consequences to students of so-called no pass/no play laws and regulations that exist in some states, the need for teachers to be highly competent in explaining their grading procedures becomes an important consideration. (In Texas, for example, students who wish to participate in extracurricular activities are required to maintain a C average—that is, at least a 70—in each subject during a grading period.)

The professional development issues concerning the creation of methods of alternative assessment and the interpretation of assessment information are directly linked to those concerning effective communication of results. In fact, if teachers have given careful thought to issues of assessment development and interpretation, then they already have the means to explain the results to *any* interested party. For example, in an open-ended mathematics task, the task itself, the scoring guide, and exemplar student responses for each scoring level can be the basis for any written or oral communication involving results. An appropriate topic for professional development activities, then, could involve explaining results to specific individuals or groups and creating an awareness that these explanations may likely be quite different, depending on the audience. For example, during a workshop a teacher who has used student journals could tell fellow teachers (1) how this assessment tool was explained to parents during the school's annual "open house," and (2) how it was presented to the district mathematics curriculum coordinator, emphasizing the similarities and differences between the two explanations.

Careful planning for, and the implementation of, alternative assessments can also help teachers address the other communication responsibility—addressing challenges to their evaluation methods. In addition to the consequences to students mentioned previously ("no pass, no play"), legal consequences to teachers and school systems could directly result from dissatisfaction with some grading processes, especially if alternative assessment methods are involved. Therefore, the need for teachers to be highly competent in defending the validity of their grading procedures becomes an important consideration. Professional development activities toward that goal would seem to be highly desirable if not totally necessary.

Because most experienced teachers have developed an awareness of ethical matters and legal responsibilities in general and have had successful experiences in justifying their own assessments methods, professional development activities might emphasize the special demands placed on teachers who use alternative assessment methods. A suggestion for such an activity focuses on role playing, in which one teacher takes the position of a hostile parent and another plays the role of the teacher who must explain and defend a grade that was based, at least in part, on assessment methods that were unfamiliar to the parent. A teacher who had previously dealt with this type of situation could critique the "performance"; the other teachers in the "audience" could offer reactions and suggestions. Role playing and other relevant activities can help teachers develop appropriate terminology and language for explaining their assessment methods and the resulting grades, thus encouraging teachers to attend to the need for effective communication of assessment results.

CONCLUSION

Ensuring that classroom teachers of mathematics can function as effective assessors will become more crucial as the vision of the NCTM *Standards* permeates teaching and learning at all grade levels. This article has presented some issues regarding the need for professional development for teachers in the area of classroom assessment and has suggested a number of relevant activities. Although there are many possible scenarios for establishing the link between professional development and classroom assessment, perhaps the best one involves "teachers helping teachers." As teachers develop the strategies and skills necessary for becoming successful mathematics assessors, it is important that they share their knowledge with their fellow teachers. One single teacher who becomes "hooked" on using effective methods of alternative assessment in the classroom can serve as an invaluable resource to other teachers in the school.

It is apropos to end this article by showcasing the efforts of a mathematics teacher who, through professional development efforts that included taking a course in alternative mathematics assessment, reading articles and reports, and trying out new ideas in her own classroom, is well on her way to assuming the role of assessor. Her own words express her feelings, expectations, and goals (Schloemer 1991, p. 13):

> I was eager to try some of the new assessment techniques that I had read about; and using them, I think that I improved my knowledge about my students' learning. My goal was to find some ways to align assessment with the *Standards* ... and I feel that I have met this goal. I now have some usable techniques to evaluate student communication, problem solving, reasoning, mathematical power, and mathematical disposition. I wondered how I could ever implement the *Standards*. Now I have some techniques that work for me.

As she continues to develop and use alternative assessment methods in her mathematics classroom, this teacher and others like her can become model assessors to their fellow teachers through sharing scoring guides, detailing what worked (and what didn't), and showing samples of student work. In a more formal mode, they can conduct professional development workshops in their schools or for their districts or give presentations on classroom assessment at state, regional, and national conferences. By realizing that "now I have some [assessment] techniques that work for me," this teacher and teachers like her have the potential to begin a grassroots movement that can enable other teachers to gain the assessment skills so necessary for the mathematics classroom as envisioned in the NCTM's *Curriculum and Evaluation Standards* and *Professional Teaching Standards*.

REFERENCES

Archbald, Doug A., and Fred M. Newmann. *Beyond Standardized Testing: Assessing Authentic Academic Achievement in the Secondary School.* Reston, Va.: National Association of Secondary School Principals, 1988.

Baron, Joan B. "Performance Assessment: Blurring the Edges among Assessment, Curriculum and Instruction." In *Assessment in the Service of Instruction,* edited by Audrey B. Champagne, Barbara E. Lovitts, and Betty Calinger, pp. 127–48. Washington, D.C.: American Association for the Advancement of Science, 1990.

Cain, Ralph W., and Patricia A. Kenney. "A Joint Vision for Classroom Assessment." *Mathematics Teacher* 85 (November 1992): 612–15.

California State Department of Education. *A Question of Thinking.* Sacramento: The Department, 1989.

Charles, Randall, Frank Lester, and Phares O'Daffer. *How to Evaluate Progress in Problem Solving.* Reston, Va.: National Council of Teachers of Mathematics, 1987.

Clark, John L. "The Toronto Board of Education's Benchmarks in Mathematics." *Arithmetic Teacher* 39 (February 1992): 51–55.

Clarke, David J. *Assessment Alternatives in Mathematics.* Canberra, Australia: Curriculum Development Centre, 1988.

Clarke, David J., Doug M. Clarke, and Charles J. Lovitt. "Changes in Mathematics Teaching Call for Assessment Alternatives." In *Teaching and Learning Mathematics in the 1990s,* 1990 Yearbook of the National Council of Teachers of Mathematics, edited by Thomas J. Cooney, pp. 118–29. Reston, Va.: The Council, 1990.

Elliott, Portia C., ed. "Assessment." Focus issue of the *Arithmetic Teacher,* February 1992.

Goehring, Harvey J., Jr. "Course Competencies for Undergraduate Courses in Educational Tests and Measurement." *Teacher Educator* 9 (1973): 11–20.

Gullickson, Arlen R. "Teacher Education and Teacher-Perceived Needs in Educational Measurement and Evaluation." *Journal of Educational Measurement* 23 (1986): 347–54.

Kroll, Diana Lambdin, Joanna O. Masingila, and Sue Tinsley Mau. "Cooperative Problem Solving: But What about Grading?" *Arithmetic Teacher* 39 (February 1992): 17–23.

Lester, Frank K., Jr., and Diana Lambdin Kroll. "Implementing the Standards: Evaluation: A New Vision." *Mathematics Teacher* 84 (April 1991): 276–84.

Linn, Robert L. "Essentials of Student Assessment: From Accountability to Instructional Aid." *Teachers College Record* 91 (Spring 1990): 422–36.

Mumme, Judy. *Portfolio Assessment in Mathematics.* Santa Barbara: California Mathematics Project, University of California, Santa Barbara, 1990.

National Council of Teachers of Mathematics. *Curriculum and Evaluation Standards for School Mathematics.* Reston, Va.: The Council, 1989.

———. *Professional Standards for Teaching Mathematics.* Reston, Va.: The Council, 1991.

Pandey, Tej. *A Sampler of Mathematics Assessment.* Sacramento: California Department of Education, 1991.

Resnick, Lauren B., and Daniel P. Resnick. "Assessing the Thinking Curriculum: New Tools for Educational Reform." In *Changing Assessments: Alternative Views of Aptitude, Achievement and Instruction,* edited by B. R. Gifford and M. C. O'Connor, pp. 35–75. Boston: Kluwer Academic Publishers, 1992.

Schloemer, Cathy G. "Standards-Aligned Assessment in Precalculus." Mimeographed. Pittsburgh: Division of Mathematics and Science Education, College of Education, University of Pittsburgh, 1991.

Shell Centre for Mathematical Education. *Problems with Patterns and Numbers.* Nottingham, England: University of Nottingham, 1984.

Shepard, Lorrie A. "Why We Need Better Assessments." *Educational Leadership* 46 (April 1989): 4–9.

Stenmark, Jean Kerr, ed. *Mathematics Assessment: Myths, Models, Good Questions, and Practical Suggestions.* Reston, Va.: National Council of Teachers of Mathematics, 1991.

Vermont Department of Education. *Looking beyond "The Answer": The Report of Vermont's Mathematics Portfolio Assessment Program.* Montpelier, Vt.: The Department, 1991.

Webb, Norman, ed. *Assessment in the Mathematics Classroom, K–12.* 1993 Yearbook of the National Council of Teachers of Mathematics. Reston, Va.: The Council, 1993.

Webb, Norman, and Diane Briars. "Assessment in Mathematics Classrooms, K–8." In *Teaching and Learning Mathematics in the 1990s,* 1990 Yearbook of the National Council of Teachers of Mathematics, edited by Thomas J. Cooney, pp. 108–17. Reston, Va.: The Council, 1990.

10

A Multimedia Approach to the Professional Development of Teachers: A Virtual Classroom

Mary M. Hatfield
Gary G. Bitter

C ONSIDER the following scenario:

> A university methods professor is teaching an elementary/middle school mathematics methods class and discussing the role of discourse in the mathematics classroom. The instructor selects a video clip from the videodisc and shows a first-grade teacher asking the students to predict how many square tiles it will take to cover the shape on the geoboard. After the children offer various predictions, the teacher says, "Try it. Let's find out!" The children test their predictions, then discuss how their predictions compare to their finding. The teacher poses more questions about covering to determine area. In a later scene, the children use only geobands to enclose the area. Again the task is to predict how many tiles will cover the space. One child is reluctant to give a prediction. The teacher asks probing questions using other numbers as referents: "Do you think it takes fewer than four?" The student shakes her head. "More than ten?" The student nods her head. "Why do you think so?" The university professor stops the videodisc to discuss with the prospective teachers many aspects of the teacher's role in discourse: posing questions to challenge students' thinking, listening to students' ideas, deciding when to clarify an issue and when to model, and asking students to justify their ideas orally. Several video scenes are shown again for further observation and discussion.

This scene illustrates instances of the decisions that teachers frequently face and behaviors they use while teaching students, such as deciding when to clarify an issue, listening to students' ideas and justifications, determining students' conceptual understandings, deciding consequences for classroom behavior, and providing encouragement. The decisions teachers make are part of what Copeland (1989) calls "the teacher's clinical reasoning" (p.10). The quality of teachers' clinical reasoning, or thought processes by which teachers decide what teaching behaviors to employ, is part of the research (Berliner et al. 1988; Carter and Doyle 1986) on novice/expert teachers. It is a concern that embraces "a perspective derived more from cognitive than from behavioral psychology in that it focuses on teachers' *understanding* of classrooms" (Copeland 1989, p. 10). After viewing these video scenes, prospective teachers can discuss the implications about the teaching and learning of mathematics, analyze the teacher's clinical reasoning, and review important instructional events. Teacher educators can help them judge the appropriateness of representations in teaching and determine what conceptual understandings are constructed by the students as the preservice teachers look beyond the surface level. "Preservice teachers need a chance to see and think about other ways of finding out what pupils know and understand, particularly ways that allow them to explore the dynamics of the process by which pupils make sense out of the world" (McDiarmid, Ball, and Anderson 1989, p. 201).

BELIEFS AND ASSUMPTIONS ABOUT TEACHING

The challenge of teacher education programs is to build meaningful experiences that develop frameworks for thinking about teaching; to help prospective teachers observe and interpret various classroom practices, procedures, and processes; and to encourage prospective teachers to question their assumptions and beliefs about the teaching and learning of mathematics. As Schmidt and Kennedy (1990) suggest, "If reformers want to improve the content and pedagogy of teaching, they need to confront teachers' prior beliefs. Providing new curricula, new incentives, or new regulations is not likely to significantly alter teaching practices if teachers do not understand or do not agree with the goals and strategies implicit in these new devices" (p. 2).

Research by McDiarmid, Ball, and Anderson (1989) on the importance of teacher beliefs states, "Students typically begin their teacher education programs with the view that teaching is telling and learning is accruing information. Years of their 'apprenticeship of observation' (Lortie 1975) and of watching the public behaviors of teachers have led them to believe that they understand what is required for teaching" (p. 201). Along with this "absorption model" of the teaching-learning process, many prospective teachers view mathematics as a subject governed by rigid rules, algorithms, and procedures with the major emphasis on obtaining right answers. The mathematics curriculum encountered by prospective teachers may have emphasized computational facility, speed, and accuracy. Mathematical modeling, cooperative problem solving, and other components of the National Council of Teachers of Mathematics (NCTM) *Curriculum and Evaluation Standards for School Mathematics* (1989) have traditionally not been part of their learning experiences. Prospective teachers need to acquire more flexible understandings of mathematics and be able to consider the appropriateness of the textbook, the models, and materials for the teaching and learning of mathematics.

Teacher education should take into consideration preservice teachers' prior conceptions of learning, teaching, subject matter, and learners of mathematics. Ideas about learning have shifted away from the teacher as a disseminator of information and toward the learner as the center of active learning. The learner is responsible for constructing understanding and seeing relationships by engaging in hands-on activities that encourage mathematical thinking. These conceptions of learning support the need for "teacher educators to design teacher education programs that foster the development of schemata for teaching—that is, learning how to teach occurs in ways that are similar to other learning" (Barnes 1989, p. 14). Prospective teachers must experience learning in a context that encourages the development of mathematical thinking as they construct new understandings of mathematics. Discussions with prospective teachers give evidence of images of teaching and the teacher's role in the teaching-learning process that do not match this new perspective.

LINKING CONTENT AND PEDAGOGY

A major issue in improving teacher education in the United States is providing situations in which prospective teachers encounter the link between the subject matter and pedagogy, that is, between the content of the subject and how it should be taught. Preservice teachers need to know *more* than subject matter. They should be given not only instructional techniques for handling specific mathematical topics but also techniques for understanding how children learn the content and possible misconceptions students might make. Calderhead (1989) describes how in initial periods of classroom observations student-teachers have difficulty understanding the classroom processes because "they lack the concepts with which to perceive what is going on in classrooms ... and need guidance to learn to discriminate the noise and activity of classroom life" (p. 47). In most instances the situation of the university classroom does not allow the link to be strengthened through actual observations of classroom experiences. When prospective teachers are given information about effective teaching, the information fails to be transformed into meaningful data to be applied in an actual teaching situation.

EFFECTIVENESS OF MULTIMEDIA-BASED INSTRUCTION

Multimedia instruction has been effective for training in industry, medicine, and the military since around 1975. Learners using interactive video display positive attitudes toward learning in this manner and tend to learn as well as or better than when using traditional instruction. Multimedia instruction is a powerful tool for education, since this technology can combine audio, text, animation, music, graphics, photos, print materials, and full-motion video to produce a simulated learning and teaching environment with the user actively engaged.

Multimedia instruction is a relatively new concept that combines videodisc technology and a digital computer. Essentially, this technology allows audiovisual materials to be viewed in ways specific to the viewer. Originally, interactive video (in the form of videotape) was used as a medium to transmit various clips of content material. This created the "ultimate VCR" (Amthor 1992), wherein a specific database was instantly accessible at the whim of a user. With the advent of the videodisc, a user could access various segments within a particular database and view them in any order (as opposed to the linear progression that was specified with a videotape). The order, speed, and progression of the database contained on the videodisc is completely dependent on a user's input through the computer. This technology made possible a multitude of engaging applications—from video-based training to education and entertainment.

This new technology has not been without problems. Many people who face this frontier have limited computer skills and are intimidated by the new technology. However, given that new information to be learned is growing at an exponential rate each day, it is imperative that some medium be available for storing and disseminating information to those who need to know. A computer is the best-known tool to accomplish this task. Data can be held and disseminated in two ways. One way is by a fixed database where a tape or disk contains data points to be accessed, viewed, or studied by a viewer. This is a "conventional" method, and the data are useful until they become obsolete. The content cannot be updated until it is replaced by a completely new videodisc or tape. A second kind of storage medium for computerized videodisc technology is that in which permanent data are kept on a disk, but new data can be processed and stored for future use. The capabilities for processing information create a self-sustaining, self-regulating system that updates old information and augments long-standing programs and data files.

This innovation for an adaptable and programmable videodisc has been termed DVI (Digital Video Interactive) multimedia technology. Such technology allows the delivery of a truly integrated VCR-quality video and audio in an environment of all digital computing with the capability for networking, novel categories of exploratory applications, and traditional applications in a multimedia format (Amthor 1992). Such technology appears to be the proverbial "road map" into the future.

Accessing and using such technology requires knowing how to perform certain commands in a specified, fairly standard sequence to observe and manipulate the data. With practice, these "rituals" become so routine that the database or application can be directly "touched." Thornburg (1992) suggests that interfacing with a computer requires a program that allows the user to "step through" the program and treat the task much as if physically manipulating it. The interface is a "mask"; good software, with reasonable rates of feedback and a variety of options and responses, provides virtual reality in performing a task. Technical jargon, limited activities, arcane commands, slow processing, and limited memory capacities disrupt the flow of thought and interfere with the user's creativity and learning. Thus, a user-friendly computer interface for accessing the database of any videodisc is essential for effective understanding of the material contained on the videodisc and for effective use of the technology. Figure 10.1 shows the common characteristics of interactive multimedia technology.

Characteristics of Interactive Multimedia Technology

1. *Promotes active versus passive learning.* Interactive multimedia technology requires a learner to actively interface with an application using commands to produce the desired results and process the responses given by the computer.

2. *Offers models or examples of exemplary and nonexemplary instruction.* A learner can view scenarios of both adaptive and nonadaptive methods or examples that they are interested in emulating or avoiding in the future.

3. *Is illustrative and interactive.* Multimedia technology provides the learner with many visual stimuli to facilitate the comprehension of a concept or method. Concepts, images, and ideas are available for the user to interact with or manipulate.

4. *Facilitates the development of decision-making and problem-solving abilities.* By using a particular application, a user can acquire specific and general problem-solving skills that can be transferred to other situations and settings. Learners with more information about a particular topic can more easily make better decisions about problems they encounter.

5. *Provides user control and multiple pathways for accessing information.* A learner is completely autonomous at controlling a particular program's application and can access information in a variety of modes depending on the learner's learning style and preference for viewing a program.

6. *Provides motivation and allows for the variability of learning styles.* The design of multimedia technology is flexible in that it allows a multitude of methods that are conducive to individual learning styles. This characteristic encourages the learner to be comfortable and promotes learning.

7. *Facilitates the development of perceptual and interpretational abilities.* The nature of interactive multimedia technology provides a learner with information about the material being viewed. The design permits self-analysis and points out important nuances of which a learner needs to be aware for exemplary emulation of the concept being studied.

8. *Offers efficient management of time for learning and less instructional training time.* Multimedia technology supplies an array of information. Thus, learning a concept, skill, or method requires less time than learning by conventional methods.

9. *Allows for numerous data types.* Multimedia technology allows for the presentation of subject matter and content in a variety of forms (e.g., animation, graphics, voice, text, motion video) that can be used by the learner. Techniques, instructional contexts, and social and motivational factors to facilitate the progression of the learner through a program are contained in intact databases for a user.

10. *Offers multilingual presentation.* The technology exists whereby programs and units can be presented in different languages.

Fig. 10.1

USING TECHNOLOGY IN PRESERVICE EDUCATION

Teacher educators are always seeking ways to bring the reality of the elementary school classroom to the university setting along with enhancing the prospective teachers' awareness of the teaching-learning process. Technology can make a major contribution to this aspect of the reform needed in teacher education. Interactive multimedia technology serves this instructional need by providing video episodes of teaching that allow preservice teachers to analyze the teaching, pose alternative teaching strategies, and gain confidence in learning about different teaching situations. "New developments in videodisc and computer technologies make it possible to combine traditional text-based information about teaching and learning with real class incidents to create information-rich 'cases' for prospective teachers to analyze" (Goldman and Barron 1990, p. 26). From the viewpoint of industrial design, technology is not as important as the instructional setting in which it is embedded (Bitter and Hatfield 1992a). Teacher educators need to use the technology as a mode of classroom interaction. As Brooks and Kopp (1989) assert, "If teacher education is to meet its responsibility to prepare teachers for the information age, then teacher educators have a professional responsibility to provide leadership in developing the full potential of existing and emergent technologies in teacher training" (p. 2).

To help develop competencies required for beginning teachers, the interactive multimedia technology allows novice teachers to see, explore, and evaluate the teaching and learning of mathematics. The complex, multifaceted act of teaching takes place in a classroom where several events typically occur simultaneously. Using interactive multimedia instruction, methods instructors can select an appropriate situation or classroom event and ask their university students to identify, analyze, and define certain situations and principles that apply to that event. A video segment can be viewed numerous times for multiple analyses from different perspectives. As Barnes (1989) states, "Teachers must understand all the factors that interact in a teaching situation—learners, learning, subject matter, and teaching—if they are to make sound judgments.... Based on knowledge gained from reflection on and critical analysis of previous experiences, professionals can postulate alternative interpretations of a situation and ways to respond. They can reflect and revise their initial understandings" (pp. 16, 19).

USING TECHNOLOGY IN TEACHER ENHANCEMENT IN-SERVICE PROGRAMS

The application possibilities for using technology in teacher enhancement in-service programs are constantly changing in both magnitude and form. The body of research on the application of technology to the training of teachers indicates that technology does not hold a secure place in the professional development of teachers. As Brooks and Kopp (1989) acclaim, "If first-year teachers are expected to be creative and facile with technology, they deserve systematic exposure to technological enhancements at all levels of a coherent, interrelated preservice curriculum. Entry year teacher competence with technology can be achieved with (*a*) coordinated university faculty research on the applications of technology to teacher training and (*b*) the university faculty instructional modeling of technological enhancements throughout the preservice curriculum" (p. 4).

Evaluation and analysis of teaching are goals for experienced teachers as well as prospective teachers. The NCTM *Professional Standards for Teaching Mathematics* (1991) calls for teachers to analyze multiple episodes of classroom teaching. Multimedia technology can provide experienced teachers with opportunities to engage in reflective teaching, to focus on ways of presenting particular content in a range of representations, and monitor learner outcomes. Integrating technology into classroom practice is a complex process of professional development. Teachers must gain confidence with the equipment, find techniques for using the technology in meaningful experiences, investigate learner competencies, and plan new forms of classroom organization. In-service education must model the kind of teaching and learning that is desired in the classroom application. Therefore, professional development leaders should give teachers the opportunity to construct their own knowledge and to apply this knowledge. This assertion implies that teachers should engage in extensive hands-on activities with multimedia technology to practice their skills. Multimedia instruction allows users to regulate their own learning experiences and to practice in simulated settings that focus on the targeted knowledge and skills without having to deal with the exigencies of real classrooms. Alternative assessment techniques can be explored, settings can be manipulated, and feedback can be supplied. Interactive multimedia instruction offers a way to analyze the cognitive strategies that teachers use and to rescript lesson plans and activities.

Examples of Interactive Multimedia Projects

The following projects illustrate how technology can make a contribution to the type of reform needed in preservice teacher education programs. The three examples of using interactive multimedia instruction in mathematics methods courses demonstrate ways to encourage prospective teachers to question their beliefs about teaching, to connect their subject-matter knowledge to actions, and to monitor their understanding of the teaching and learning of mathematics.

To help mathematics methods students apply the theory encountered in their college courses to real classroom settings, Goldman and her colleagues at Peabody College of Vanderbilt University developed a hypermedia-based, interactive videodisc system to supply prospective mathematics teachers with a context for exemplary elementary school classroom teaching (Goldman and Barron 1990). The design of the videodisc materials embodies the assumption that prospective teachers need knowledge about effective mathematics teaching—knowledge that is best acquired if information is encountered in context. The videotaped examples from real teaching situations furnish a context for analyzing teaching as essentially a problem-solving endeavor and offer the prospective teacher keen analytical skills. Since multimedia technology allows for the integration of a host of media (e.g., video, text, digitized pictures), it provides accounts of classroom instruction for analysis and discussion.

A videodisc with HyperCard control allows the viewer or course instructor to analyze the same mathematical or science lessons taught and managed by both a novice teacher and an experienced teacher. Goldman and her associates analyzed videotapes of whole-class lessons and clinical interviews with children and edited them to produce videodiscs. Video segments linked to research on teaching and learning mathematics or to curriculum and teaching guidelines are analyzed by prospective teachers. These analyses force them to examine their beliefs and assumptions about teaching mathematics as well as examine the instructional approach, teachers' directions and demonstrations, students' reactions to instructions, classroom management, and concept and symbol development. Group comparisons of preservice teachers taught (1) with conventional instruction and (2) with conventional instruction plus the HyperCard system revealed that the use of hypermedia materials influences the way preservice teachers conduct mathematics lessons in their practicum associated with their methods class. These differences, however, decrease when comparisons are made one to three semesters later during student teaching.

At Michigan State University, another project funded by NSF, "Mathematics and Teaching through HyperMedia" (Lampert and Ball 1990), is designed for teacher educators and preservice students of teaching. The goal is to furnish a system that has the potential to alter the way prospective teachers think about mathematics and the teaching of mathematics. The project created and field-tested a collection of multimedia tools for exploring and constructing knowledge about mathematics teaching and learning in elementary schools. The project was constructed using a hypermedia system that links multiple sources of information about the learning and teaching of mathematics whereby audio and video records of mathematics lessons were collected during one year. Also, written work done by students, interviews with students about skills and attitudes toward mathematics, teachers' daily journals, observers' field notes, and conceptual-theoretical annotations of lessons by various professionals (e.g., mathematicians, psychologists, and sociologists) were included. The system is manipulated to catalog and assemble the collected material so that teacher educators and preservice teachers can use them to explore teaching and learning. Hypermedia technology is used to establish multiple pathways between and among data in order to construct a medium whereby a learner can develop cognitive flexibility and transfer of learning. This project positively influences novice teachers in their training to be exemplary teachers of mathematics, and the system that is developed can change the nature of the interaction that takes place between the researchers and the practitioners concerning their perceptions of the classroom environment and what is needed to improve it.

A project at Arizona State University (Bitter and Hatfield 1992b) combines technology, teacher education, and manipulatives for mathematics instruction to overcome some of the difficulties of the professional development of teachers. "Teaching Mathematics Methods Using Interactive Videodisc" (TMMUIV) is an application of multimedia technology to train teachers to use manipulatives in teaching elementary school mathematics. TMMUIV, funded by NSF, developed two units for using manipulatives to teach mathematics. These units on base-ten blocks and geoboards contain actual classroom videos of grades 1, 3, and 5, as well as university presentation segments. TMMUIV has three instructional components: (1) a video database that gives the option of selecting the video clips from a video library subdivided by fields; (2) an instructional presentation mode that allows a mathematics methods instructor or provider of professional development to use a presentation including objectives, activities, and related video clips; and (3) a stand-alone instructional system where the mathematics methods student or in-service teacher progresses through an individualized learning program. The video database furnishes a setting and capability that can be described as a virtual classroom. The database contains options for exploring classroom and university teaching and learning through many fields, depending on the goals of the instructor or presenter. The virtual classroom allows for realistic multiple observations, review, discussion, and detailed exploration. The methods instructors can select one or more fields that they want to develop for the prospective teachers' understanding. Specifically, each of the two databases—geoboards and base-ten blocks—is organized into fields on content, types of learning, manipulatives, teaching methods, levels of reasoning, and the NCTM *Professional Standards for Teaching Mathematics* (1991). Each of the fields is divided into several subfields, as shown in figure 10.2.

The TMMUIV program at Arizona State University is a point-and-click computer-controlled system that combines full-motion video, animation, graphics, text, and voice on one computer. The computer display can be projected with a color LCD panel. TMMUIV uses Microsoft Windows with multimedia extensions and is icon-driven. The icons allow users to access any components or activities by pointing and clicking on the icon of their choice. The information is stored in a video database that allows the user to use only activities that match their instructional goals.

To use the virtual-classroom concept in practice, the university instructor or in-service leader selects the desired field from the computer menu, then chooses the

Fields and Subfields for TMMUIV

Base-Ten Blocks

Manipulatives

Place-value pegboard
Tens organizers
Spinners
Base-ten blocks
Dice
Cat-face cutouts
Game boards
Hundreds organizers
Thousands organizers

Content

Addition
Equivalency
Definitions
Division
Levels of representation
Modeling
Multiplication
Proportional representations
Subtraction
Trading

Teaching Methods

Free exploration
Teaching game
Interactive lecture
One-on-one interaction
Guided discussion
Peer teaching

Levels of Reasoning

High-order convergent
Low-order convergent
Low-order divergent

Geoboards

Manipulatives

Geoboards
Square tiles
Posterboard shapes
Shapes
Geobands

Content

Area
Perimeter
Congruency
Coordinate geometry
Pentominos
Symmetry
Similarity
Reflections-rotations-flips
Unit of measure
Visualization skills

Teaching Methods

Directed performance
Guided discussion
Problem solving
Teacher directed
Worksheets

Levels of Reasoning

High-order convergent
High-order divergent
Low-order divergent
Low-order convergent

Fig. 10.2

subfield and reviews information about the video segments. Each video segment includes a description of the teaching scenario; the length of the video segment; the names, ethnicity, and gender of the students; teacher questioning methods; related research; activities; enrichment ideas; and pertinent statements from the NCTM *Professional Standards for Teaching Mathematics* (1991).

The second component of TMMUIV is a model set of lessons for classroom presentation by a provider of professional development. This multimedia presentation includes video clips from the video database, overview screens to introduce the lesson and its objectives, classroom display screens, and dialogue screens for discussion. The presentation is controlled using icons, so that the presenter can select content, video clips, activities, and dialogue to meet the instructional goals. In addition, a text and graphics tablet can be accessed to present notes, drawings, or directions throughout the lesson.

Through the individual use of the third component of the project, the laboratory system, the instructor can make assignments asking students to analyze teaching methods, view contrasting conceptions about the nature of teaching, and explore how learners are interpreting information or what misconceptions may be occurring.

A desired outcome of the project was to increase the likelihood that preservice teachers will feel competent and confident about teaching with the manipulatives. An analysis of the findings indicates that teacher education students in the TMMUIV group scored higher than those in the control group on each measure of cognitive gain, preparedness to teach with geoboards, and motivation to teach with geoboards. One of the most powerful contributions the interactive multimedia experience can make to the training of prospective teachers is the enhanced observational power that usually appears after years of teaching. Given video segments of mathematics lessons, preservice teachers in the TMMUIV group were better able to analyze instructional interactions. The group viewing the classroom video episodes reported significantly higher increases in knowledge about the reality of the elementary school classroom than the other group (Bitter 1992; Bitter and Hatfield 1993). Qualitative data were collected using open-ended questionnaires and interviews. The most recurring statements were associated with the reality of the classroom and an awareness of actual applications.

FUTURE APPLICATIONS

Consider this scenario:

The preservice teacher returns from the fourth-grade classroom with a digital videodisc data disk of his lesson. He watches the lesson in an interactive multimedia-based learning environment. Viewing his lesson with the "expert teacher" by means of the interactive multimedia computer, the novice stops the lesson and asks for an analysis of his teaching. The computer responds with questions about the degree of understanding exemplified by the students' responses during the lesson. As the lesson proceeds, the prospective teacher stops at different points to notice the degree of elaboration, any off-task behavior, and any misconceptions. After the lesson is reviewed, the novice watches a multimedia simulation of the same lesson from an expert teacher's point of view. At any point, the novice may interact with the lesson from a variety of viewpoints: directed questions to the teacher, questions or praise to the students, or probing for misunderstandings and changing the direction of the lesson. The clinical reasoning used by the expert teacher may be challenged at any point during the simulated lesson. After careful reflection and comparison, the preservice teacher's clinical reasoning is analyzed and modifications are noted. The next assignment is to develop assessment techniques to accompany the lessons just viewed. Other professionals can review the novice's lesson and supply on-line portfolio comments for later analysis.

SUMMARY

Teaching is a complex, multidimensional process that involves skills, knowledge, beliefs, judgments, and dynamic interactions. The role of prior experiences in the development of teachers' perspectives cannot be overlooked. "Since teachers teach much as they were taught, university courses for prospective teachers must exemplify the highest standards for instruction" (National Research Council 1989, p. 65). Preservice teachers need opportunities to construct their own knowledge, acquire new models of teaching, and analyze the teaching-learning process. Teacher education "requires substantial changes in the philosophy and strategies of mathematics and mathematics education instructors at the college level and beyond who are involved in the preservice and continuing education of teachers of mathematics. Instructors need to experiment with new tasks, tools, and modes of classroom interaction and share and model new instructional strategies" (NCTM 1991, p. 128).

Interactive multimedia technology presents itself as a powerful tool for instruction, since it combines the potential of many new information-related technologies. When such technologies as text, audio, graphics, still images, and full-motion video are mixed into a single, computer-controlled multimedia program, the possibilities for the classroom are virtually unlimited. In particular, the prospects for preparing elementary and secondary school teachers appear especially promising in light of such technology (Koneck et al. 1991). Professional development programs can profit from interactive multimedia instruction by providing structured observations of students and expert teachers in their classrooms. Being able to observe teaching skills and concepts in action can greatly enhance the acquisition of basic instructional skills by novice teachers as well as perfect the teaching practices of experienced teachers. Analyzing important instructional events helps develop cognitive frameworks for thinking about the teaching and learning of mathematics.

One approach to experimenting with new tools and modes of classroom interaction is the use of interactive multimedia instruction in elementary and middle school classrooms. Interactive multimedia technology should be integrated into instructional settings adapted to the needs of learners and instructors that will facilitate the transfer of complex knowledge to new situations. Multimedia instruction allows a topic to be explored in multiple dimensions in a virtual classroom using various themes and approaches that involve higher-order thinking. At a time when American education and educators are under attack for their apparent inability to produce academically competitive students, the integration of interactive multimedia instruction into our schools seems to be a step in the right direction.

REFERENCES

Amthor, Geoffrey R. "DVI Technology: The Digital Future of Multimedia." *AV Video* 14 (1992): 17–31.

Barnes, Henrietta. "Structuring Knowledge for Beginning Teaching." In *Knowledge Base for the Beginning Teacher,* edited by Maynard C. Reynolds. New York: Pergamon Press, 1989.

Berliner, David C., Pamela Stein, Donna Sabers, Pamela Brown Clarridge, Katherine Cushing, and Stefinee Pinnegar. "Implications of Research on Pedagogical Expertise and Experience for Mathematics Teaching." In *Perspectives on Research on Effective Mathematics Teaching,* edited by Douglas A. Grouws, Thomas J. Cooney, and Douglas Jones, pp. 67–95. Reston, Va.: National Council of Teachers of Mathematics, and Hillsdale N.J.: Lawrence Erlbaum Associates, 1988.

Bitter, Gary G. *Teaching Mathematics Methods Using Interactive Videodisc.* NSF Final Report. Tempe: Arizona State University, Technology Based Learning and Research, 1992.

Bitter, Gary G., and Mary M. Hatfield. "Implementing Calculators in Middle School Mathematics: Impact on Teaching and Learning." In *Calculators in Mathematics Education,* 1992 Yearbook of the National Council of Teachers of Mathematics, edited by James T. Fey, pp. 200–207. Reston, Va.: The Council, 1992a.

———. *Teaching Mathematics Methods Using Interactive Multimedia: The TMMUIV Research Results.* Monograph no. 5. Tempe: Arizona State University, Technology Based Learning and Research, 1993.

———. *Teaching Mathematics Methods Using Interactive Videodisc: The TMMUIV System.* Monograph no. 4. Tempe: Arizona State University, Technology Based Learning and Research, 1992b.

Brooks, Douglas M., and Thomas W. Kopp. "Technology and Teacher Education." In *Handbook of Research on Teacher Education,* edited by W. Robert Houston. New York: Macmillan, 1990.

———. "Technology in Teacher Education." *Journal of Teacher Education* 40 (1989): 2–8.

Calderhead, James. "Reflective Teaching and Reflective Education." *Teaching and Teacher Education* 5 (1989): 43–51.

Carter, Kathy, and Walter Doyle. "Teachers' Knowledge Structures and Comprehension Processes." In *Exploring Teachers' Thinking,* edited by James Calderhead. London: Holt, Rinehart & Winston, 1986.

Copeland, Willis D. "Technology-Mediated Laboratory Experiences and the Development of Clinical Reasoning in Novice Teachers." *Journal of Teacher Education* 40 (1989): 10–17.

Goldman, Elizabeth, and Linda Barron. "Using Hypermedia to Improve the Preparation of Elementary Teachers." *Journal of Teacher Education* 41 (1990): 21–31.

Koneck, Jeff, Neal Grangenett, Ray Ziebarth, Mary Laura Farnham, Jody McQuillan, and Becky Larson. "Factors Related to the Anticipated Future Use of Multimedia by Pre-Service Teachers." *Journal of Hypermedia and Multimedia Studies* 1 (Summer 1991): 11–18.

Lampert, Magdalene, and Deborah L. Ball. *Using Hypermedia Technology to Support a New Pedagogy of Teacher Education.* Issue Paper 90-5. East Lansing, Mich.: Michigan State University, National Center for Research on Teacher Education, 1990.

Lortie, Dan C. *Schoolteacher: A Sociological Study.* Chicago: University of Chicago Press, 1975.

McDiarmid, G. Williamson, Deborah L. Ball, and Charles W. Anderson. "Why Staying One Chapter Ahead Doesn't Really Work: Subject-Specific Pedagogy." In *Knowledge Base for the Beginning Teacher,* edited by Maynard C. Reynolds. New York: Pergamon Press, 1989.

National Council of Teachers of Mathematics. *Curriculum and Evaluation Standards for School Mathematics.* Reston, Va.: The Council, 1989.

———. *Professional Standards for Teaching Mathematics.* Reston, Va.: The Council, 1991.

National Research Council, Mathematical Sciences Education Board. *Everybody Counts: A Report to the Nation on the Future of Mathematics Education.* Washington, D.C.: National Academy Press, 1989.

Schmidt, William H., and Mary M. Kennedy. *Teachers' and Teacher Candidates' Beliefs about Subject Matter and about Teaching Responsibilities.* Research Report 90-4. East Lansing: Michigan State University, National Center for Research on Teacher Education, 1990.

Thornburg, David. "As We May Learn: Multimedia and the Intuitive Learner (Part 1)." *Designer's Forum* 1 (1992): 1, 8–11.

11

Part 2

Changing Preservice Teacher-Education Programs

Nancy Nesbitt Vacc
George W. Bright

AS A RESULT of the recent efforts of the National Council of Teachers of Mathematics (NCTM), standards now exist for both K–12 mathematics content (NCTM 1989) and mathematics instruction (NCTM 1991). Although much effort has been focused on preparing in-service teachers to meet these new standards, there is also a need to modify preservice teacher education programs to ensure that graduates are prepared for, and will be significant contributors to, the reform movement in mathematics education. Without modifications in preservice teacher education, we will find ourselves constantly having to "repair" teachers' backgrounds that do not fit the changing demands of the mathematics classroom.

Modifying preservice teacher education programs, however, is a complex and difficult task that cannot be viewed from a mathematics education perspective only. Although one goal is to align the backgrounds and experiences of new elementary school teachers with the expectations they will encounter when they begin full-time teaching, including being able to meet the national call for reform in mathematics education, a consideration of many factors is also required. One set of factors is current issues affecting changes in teacher education. A second set involves specific program changes that need to be implemented in order to address these issues. Finally, when program changes have been determined, procedures for evaluating the change effects need to be considered. It is the intent of this article, therefore, to provide an overview of (*a*) issues affecting changes in teacher education and subsequent questions to be addressed when modifying an existing course of study, (*b*) possible responses to the issues,

Preparation of this article was supported in part by National Science Foundation (NSF) Grant MDR-8954679 to the University of Wisconsin (UW). All opinions expressed are those of the authors and do not necessarily reflect the positions of either NSF or UW.

(*c*) ways to evaluate program changes, and (*d*) one teacher education program's revised course of study.

ISSUES AFFECTING CHANGES IN TEACHER EDUCATION

Revising existing mathematics education programs to meet new standards requires a consideration of general changes needed in teacher education. In addition to the call for reform in how mathematics is being taught and learned, other agencies have produced significant reports (e.g., *A Nation Prepared: Teachers for the 21st Century*, Carnegie Foundation [1986]; *Tomorrow's Teachers* and *Tomorrow's Schools*, Holmes Group [1990]) on the current status of education in the nation's schools and the need for reform in teacher education. The recommendations of these groups include more participation by classroom teachers in developing and implementing policies and programs for preservice teachers, the establishment of more collaboration between teacher education programs and public schools, and the preparation of teachers as researchers. State agencies, too, have mandated changes in preservice teacher education programs. For example, as recommended by the state's Task Force on the Preparation of Teachers (1986), all University of North Carolina students majoring in elementary and middle grades education are required to complete a second major (i.e., a minimum of twenty-four semester hours) in one of the arts or sciences. In addition, there is a demand for early and continuous field experiences throughout the professional education courses for all preservice teachers.

These reform initiatives influence what should or can be included in a preservice teacher education program. Therefore, determining necessary changes in existing mathematics education courses includes the consideration of the nature of field experiences, the classroom environment for these field experiences, influential external forces, the mathematics methods course, the coherence of the program of study, and the effort required to effect change.

Field Experiences

Considerable evidence shows that actual changes in individual performance may not match expectations when learners are given too little practice on a content task or are given practice that is without appropriate monitoring or feedback from someone who is more expert. It is equally important that feedback be consistent, regardless of the source (e.g., answer key or teacher input). It seems reasonable to expect that this generalization is also true for learning how to teach. If novice teachers get too little practice or if they receive inconsistent feedback from their university teacher educators and classroom supervisors, their expertise may not be adequately developed. In consideration of these concerns, questions such as the following need to be addressed:

1. When should field experiences commence in the program of study?
2. Should field experiences be continuous across the program of study?

3. How many hours a week should be required for each field experience?
4. Should there be a designated focus of study during each experience?
5. What assessment procedures should be used to evaluate the preservice teacher's progress in an internship setting?
6. What are the responsibilities of in-service teachers (i.e., on-site teacher educators) during the field experience?

Internship—the Classroom Environment

If a preservice teacher education program adopts a consistent philosophy on the way children learn (e.g., children construct knowledge, and this process is helped by frequent use of concrete objects), then that philosophy needs to be evident in the classrooms where preservice teachers complete their field experiences. Indeed, the entire school, from the principal to the teachers to the support staff, needs to work within this consistent philosophy. For this to be possible, university teacher educators most likely need to assume responsibility for initiating dialogue that ultimately might lead to the creation of a school environment that (*a*) is supportive of the preservice program and (*b*) includes in-service teachers working collaboratively with university faculty as on-site teacher educators. To effect this change within a given school, however, questions such as the following need to be addressed:

1. How can a cadre of on-site teacher educators, who share a philosophy about mathematics instruction that is consistent with the philosophy of the preservice teacher education program, be created and supported?
2. Do on-site teacher educators need to develop expertise in mathematics education that is different from the expertise needed by "regular" classroom teachers?
3. Do existing classroom teachers have appropriate backgrounds in mathematics and pedagogy to become effective on-site teacher educators?
4. Are appropriate resources available in the internship classroom for teaching mathematics?
5. Are school administrators aware and supportive of the current trends in mathematics education and the subsequent need for relevant professional development activities for their teachers?

External Forces

Both schools and universities are subject to pressures from external accreditation agencies. Although these pressures cannot be ignored, schools and universities need to work together to change those regulations that may be restrictive (e.g., required administration of standardized tests that assess basic knowledge only). Clearly, attention must be given to helping all professionals involved with preservice teacher preparation deal with pressures from external forces while at the same time providing high-quality instruction for students. This requires addressing the following questions:

1. Do the objectives of the standardized tests used in the district match the expectations of the preservice teacher education program in terms of student learning outcomes? If not, how can in-service and preservice teachers be helped to reconcile the discrepancies?

2. How can the expectations of accreditation agencies (both state and national) be incorporated into the preservice teacher education program?

3. Does the instruction given by preservice teachers meet the expectations of the district and of parents? If not, how can discrepancies be reconciled?

The Mathematics Methods Course

Because it can be expected that preservice teachers will teach what and how they are taught, it is essential that they be provided exemplary models of instruction during their preparation program—models represented by *both* university *and* on-site teacher educators. Therefore, consideration needs to be given to both the content included in a mathematics methods course and how it should be presented; that is, factors that are in keeping with the recommended NCTM *Curriculum and Evaluation Standards* and *Professional Teaching Standards*. Preservice teachers' knowledge base and beliefs about teaching and learning also affect the content of a methods course as well as the type and number of field experiences that need to be required. Developing a mathematics methods course, therefore, requires the consideration of questions similar to the following:

1. What content needs to be included in a mathematics methods course?

2. What field experiences need to be included and what is the best means of furnishing these experiences?

3. How many credit hours are appropriate for the methods course?

4. How will the NCTM *Standards* be addressed during the course?

5. What evaluation procedures should be employed to assess students' mathematical and pedagogical knowledge base?

6. How can a learning environment be established that will help preservice teachers construct their own knowledge concerning appropriate models of teaching and instructional strategies?

7. How should assessment of mathematics understanding be incorporated in the methods course?

The Coherence of the Program

This area ultimately may be the most important and require the most work in the process of changing a preservice program. We know very little about the different effects of coherent versus noncoherent programs on the quality of graduates (i.e., first-year teachers). However, current recommendations from reform groups suggest that coherence is likely to be an important aspect of improving the quality of graduates. Preservice teachers certainly learn from all who are responsible for the program, and if these individuals develop the program coherently, the cumulative effects may prove to be more than just additive. To determine this, we need to know the following:

1. How will a philosophy of instruction consistent with the NCTM *Standards* be infused across the teacher education curriculum?
2. How will all content areas be connected to one another?
3. What procedures need to be incorporated within the teacher education program to measure the cumulative effects of coherence on graduates?

The Effort Required to Effect Change

Everyone involved in reforming a preservice teacher education program will quickly find that the effort required exceeds all expectations. Taking the time to keep all parties involved and committed to the process is difficult even in the simplest of times, and it is even more difficult in these times of expectations for many different kinds of change. Specifically, consideration needs to be given to identifying the following:

1. What commitment will be required of preservice teachers?
2. What commitment will be required of in-service teachers and other school personnel?
3. What commitment will be required of teacher educators in the university or college?

POSSIBLE RESPONSES TO THE ISSUES

Attempts to answer the questions above as a first step in revising our preservice program for elementary education majors led us to recognize the value of using classroom teachers as on-site teacher educators and establishing professional development schools. In addition, we believe that including cognitively guided instruction in the mathematics methods course is one mechanism that supports the development of an instructional philosophy consistent with that of the NCTM *Standards*. Discussion concerning each of these alternatives is presented below.

On-Site Teacher Educators

Appropriate opportunities for preservice teachers to apply their theoretical and methodological understandings in classroom settings have long been a highly valued and essential component of teacher preparation programs (Shapiro and Sheehan

1986). Also, relevant field experiences critically influence the transition from preservice to in-service teaching (Freeland 1979; Gallemore 1981). Although the supervision of these field experiences has historically been provided by a university faculty member, several researchers have found that classroom teachers are more effective than university faculty in influencing preservice teachers' learning, teaching performance, and attitudes (Emans 1983; Haberman and Harris 1982; Vacc and Russell 1992). In addition, classroom teachers devote substantially more time and effort to working with preservice teachers than university supervisors do (McIntire and Morris 1980). On the basis of these findings, it seemed clear that we should incorporate on-site teacher educators into our preservice teacher education program.

Professional Development Schools

The professional development school (PDS) concept (Holmes Group 1990) involves a learning environment where everyone (i.e., students, in-service and preservice teachers, principals, and other school personnel) continues to learn. The basic premise is that schools dedicated to the education of all personnel will be better places for students to learn. The PDS model also supports sustained experiences in classrooms that help preservice teachers integrate what they are learning about teaching with what they are observing, doing, and feeling in classrooms (i.e., making connections between theory and practice). As a result of the partnership between university and on-site teacher educators, classroom teachers are in a position to (a) make significant contributions to planning, implementing, and evaluating changes in the teacher-preparation curriculum and (b) share their findings and ideas with others through professional writing, presentations at professional conferences, and both formal and informal in-service sessions. The PDS seemed to be a good mechanism for letting in-service teachers become on-site educators.

Cognitively Guided Instruction

Cognitively guided instruction (CGI) is an approach to teaching and learning mathematics through which teachers use their knowledge of students' cognitions to guide the instructional decisions they make as they try to adjust the level of content to match students' performance levels. CGI was originally developed because of the increasing knowledge base about how students in primary grades learn mathematics. The cooperative relationship between in-service teachers and teacher educators that develops during the implementation of CGI is highly consistent with the establishment of PDSs. (More information on CGI is contained in chapter 27, "Using Knowledge of Children's Thinking to Change Teaching," which focuses on the in-service education of primary-grade teachers).

EVALUATING PROGRAM CHANGE EFFECTS

In evaluating changes in preservice teacher education to meet the NCTM *Standards,* one must consider variables directly related to teaching and learning mathematics. As a result, an evaluation of the following, at a minimum, is recommended: (*a*) beliefs and knowledge about teaching and learning in general and mathematics in particular, (*b*) teaching techniques in mathematics instruction, (*c*) types, quantity, and quality of students' verbalizations during mathematics instruction, (*d*) preservice teachers' interpretations of students' performance in mathematics, and (*e*) interventions of university and school-based supervisors.

Consideration also needs to be given to conducting both summative and formative evaluation procedures. A problem exists, however, with the paucity of valid instruments to measure many of the variables cited above. The Mathematics Beliefs Scale (Fennema, Carpenter, and Loef 1990; Peterson, Fennema, Carpenter, and Loef 1989) is one of the few techniques for determining a teacher's beliefs about teaching and learning mathematics. Other sources of data include journal entries, classroom observations, videotapes of preservice teachers' lessons, and interviews.

Interviews might be structured around the guidelines of Belenky et al. (1986), which focus on an individual's "ways of knowing." Specifically, interviewees can be classified as *received knowers* (i.e., they learn from listening to others, especially to authorities who know the "truth"), *subjective knowers* (i.e., they find the source of knowledge in the self, listening to their own inner voice, recognizing that there are multiple truths that are often difficult to communicate), *procedural knowers* (i.e., they believe that knowledge is acquired, developed, and communicated through the deliberate and systematic use of procedures, which are abstract and analytic for some and narrative and holistic for others), and *constructive knowers* (i.e., they understand that knowledge is constructed and the knower is seen as shaping the known). Although interviews and observations are rich sources of data, analyzing the data they provide is time-consuming as well as subject to rater or coder bias; interpretations, therefore, need to be made with care.

A REVISED COURSE OF STUDY

The most obvious changes in our preservice program were the establishment of PDSs and the establishment of inquiry teams of preservice teachers who take all their professional courses together as a cohort. Another major modification was the inclusion of cognitively guided instruction in the mathematics methods course.

Preservice Teacher Cohorts

Two or three cohorts of undergraduate preservice teachers (approximately thirty students each) and one cohort of graduate preservice teachers (approximately twenty-five students) are enrolled each year, with each cohort assigned to two or three PDSs. Classroom teachers at each PDS serve as on-site teacher educators and meet regularly with university faculty to plan the field experiences for the preservice teachers (e.g., observations, tutoring, small- and whole-class instruction, and individual assessment of children). The on-site teacher educators also model instructional activities for the various methods courses.

Preservice teachers in the undergraduate cohorts are required to take a mathematics placement test at the beginning of their freshman year and to complete six semester hours of unspecified coursework in the Department of Mathematics before their junior year. They complete the mathematics methods course during the fall semester of their senior year. In addition to other methods courses, they complete ten hours a week of internship during both semesters of the junior year and the fall semester of the senior year. Their student-teaching experience is completed at one of the PDSs during the spring semester of the senior year.

Members of the graduate cohort of preservice teachers are enrolled in a master's degree program in elementary education, but each has a baccalaureate degree in an area other than education. The program of study is completed in one full academic year and two summers, excluding any prerequisite courses needed for certification. For example, prerequisites include six semester hours of non–education mathematics, excluding basic mathematics. As part of the program, the graduate cohort completes five hours of field experience each week during the fall semester and four and one-half days of full-time "clinical internship" each week during the spring semester.

Cognitively Guided Instruction

Being able to include CGI in our course of study was fortuitous and came about as a result of an invitation to participate in the CGI Primary Preservice Teacher Education Project, which is funded by a National Science Foundation grant to the Wisconsin Center for Education Research. This project is attempting to determine if the inclusion of CGI in preservice teacher education can improve the performance of beginning teachers at the elementary school level.

Evaluating Program Change Effects

Although we believed, on the basis of the literature, that modifying our program was essential if we were to prepare preservice teachers to meet the national call for reform in mathematics education, we were also interested in documenting the program's effects on the preservice teachers. Further, we wanted to know more about the effects of including CGI in the methods course. We believe that CGI fits very well with the PDS model. CGI and PDS are individually valuable additions to our program, but the synergy of the two amplifies the effect. Documenting this effect became important. As a result, we established a quasi-experimental, control-group, two-year study to evaluate the effects of incorporating CGI within the program. The mathematics methods course for one of the undergraduate cohorts (hereafter referred to as CGI cohort) and for the graduate cohort included CGI. The second undergraduate cohort (non-CGI cohort) was not prepared in CGI. The following discussion includes our reflection on the first three semesters of the evaluation project.

Interviews. Part of our formative and summative evaluation is to document the ways that preservice teachers change in their understanding about how knowledge, both personal and public, is constructed. We suspect that CGI probably works best with teachers who believe that students are ultimately the shapers of their own knowledge. Consequently, we wanted to know how our preservice teachers perceive the business of knowledge making. Therefore, during the first semester of the students' program, all the preservice teachers in the CGI cohort were interviewed about the acquisition of knowledge. Some were classifiable in each of the four categories of ways of knowing, but many showed evidence of being between categories; for example, they displayed remnants of the characteristics of Subjective Knowers while moving toward Procedural Knowers. We expect that by the conclusion of the study, more preservice teachers will be classifiable in the Constructed Knowers category.

Beliefs about mathematics education. Probably the most important quantitative data are the responses to the Mathematics Beliefs Scale (Fennema, Carpenter, and Loef 1990), which has four subscales: (1) Role of the Learner—children construct or receive mathematics knowledge, (2) Relationship between Skills—mathematics skills are embedded or isolated in problems, (3) Sequence of Mathematics Topics—instruction builds on children's knowledge or on the structure of mathematics, and (4) Role of the Teacher—instruction facilitates children's construction or the teacher's presentation. To date, the scale has been administered to the undergraduate cohorts at the beginning of the 1991 and 1992 fall semesters and at the beginning of the 1993 spring semester. The only comparative analysis to report here is for the initial administration. All three groups of preservice teachers scored nearly the same on the Role of the Learner subscale. For the subscale on skills and understanding, however, graduate preservice teachers scored significantly higher ($p < .05$) than either group of undergraduate preservice teachers. On each of the other subscales, the graduates' scores were higher (i.e., had a more constructivist orientation) than those of the CGI undergraduates; the non-CGI-prepared undergraduates' orientation was in the middle and did not differ significantly from either of the other groups.

Changes in beliefs can also be noted by examining each group separately. Although members of the two undergraduate cohorts did not differ in their beliefs about teaching and learning mathematics at the beginning of the study, significant differences ($p < .05$) existed between the two groups after the first year of the program for subscales 2 and 3. The non-CGI cohort had more constructivist beliefs than the CGI cohort. However, these differences no longer existed after the third semester, during which the students completed the mathematics methods course. Although an explanation for the differences at the second testing remains unclear, it appears that an effect may exist related to field experiences and differences in PDSs. Preservice teachers in the CGI cohort had field experiences in three different schools, each of which included a diverse student population. Preservice teachers in the non-CGI cohort completed field experiences in two different schools where the student populations were not as diverse.

The graduate cohort's beliefs about teaching and learning mathematics changed substantially during the course of the program; its members' mean scores on the Beliefs Scale were significantly ($p < .001$) higher at the end than at the beginning. Too, these preservice teachers were more cognitively based in their beliefs about mathematics instruction at the end of one year than the undergraduate preservice teachers were at the end of one and one-half years. These differences, which were not unexpected given the data at the beginning of the study, may be due to the differences in types of students.

In addition to the quantitative data, beliefs about teaching and learning mathematics have been or will be addressed through interviews with members of the CGI cohort conducted during the second and final semesters of the course of study and through journal entries completed during the third semester of study.

Teaching competencies. The undergraduate preservice teachers completed their third-semester internship (i.e., the fall of their senior year) in the same classroom where they are currently student teaching. Thus, continuous observations over two semesters and the videotaping of instructional activities are supplying formative and summative data concerning a preservice teacher's progress in teaching mathematics and making connections between theory and practice. In addition, on-site teacher educators are able to furnish ongoing evaluation of the preservice teachers' progress and the overall course of study. Likewise, preservice teachers are supplying formative and summative data concerning the effectiveness of the on-site teacher educators, coursework, and university instructors. It seems, from reports of the on-site teacher educators, that the CGI preservice teachers who are in classrooms where CGI is used by the on-site teacher educator may be making more progress toward implementing CGI than CGI preservice teachers who are in classrooms where CGI is not used by the on-site teacher educator. However, firm conclusions must await full examination of the data currently being gathered.

CONCLUSIONS

Several issues have evolved as part of the change process to date: (*a*) creating a cadre of on-site teacher educators who share a philosophy about mathematics instruction that is consistent with the program's philosophy, (*b*) determining adequate and appropriate course content and field experiences, (*c*) identifying and providing needed professional development activities within the PDSs, (*d*) incorporating expectations of accreditation agencies into the program, (*e*) incorporating procedures within the program to measure the cumulative effects of coherence on graduates, and (*f*) identifying the amount of commitment required of all those involved with the change process. Qualitative data provided by all participants are being analyzed to determine ways to address these issues.

Changing preservice teacher education programs is difficult because of the large number of factors that need to be considered and the lack of control of all variables. Preliminary findings from our project appear to support the possibility of changing what preservice teachers believe about teaching and learning in general and mathematics education in particular. Yet, we need further study of the effects of (*a*) field experiences in PDSs, (*b*) the use of on-site teacher educators, and (*c*) the inclusion of CGI in the methods course.

Our program continues to evolve, and not everything we try will work. However, we believe that we have greatly improved the coherence of the program, which will ultimately result in better teaching and then better learning for children.

REFERENCES

Belenky, Mary F., Blythe B. Clinchy, Nancy R. Goldberger, and Jill M. Tarule. *Women's Ways of Knowing*. New York: Basic Books, 1986.

Carnegie Forum on Education and the Economy: Task Force on Teaching as a Profession. *A Nation Prepared: Teachers for the 21st Century*. Washington, D.C.: Carnegie Foundation, 1986.

Emans, Robert. "Implementing the Knowledge Base: Redesigning the Function of Cooperating Teachers and College Supervisors." *Journal of Teacher Education* 34, (1983):14–18.

Fennema, Elizabeth, Thomas Carpenter, and Megan Loef. *Mathematics Beliefs Scales*. Madison, Wisc.: University of Wisconsin, 1990.

Freeland, Kent. "Assessing Student Teaching." *Teacher Education* 15 (1979): 11–16.

Gallemore, Sandra L. "Perceptions about the Objectives of Student Teaching." *Research Quarterly for Exercise and Sport* 52 (1981): 180–90.

Haberman, Martin, and Patricia Harris. "State Requirements for Cooperating Teachers." *Journal of Teacher Education* 33 (1982): 45–57.

Holmes Group. *Tomorrow's Schools: Principles for the Design of Professional Development Schools*. East Lansing, Mich.: The Group, 1990.

———. *Tomorrow's Teachers*. East Lansing, Mich.: The Group, 1990.

McIntire, D. John, and William R. Norris. "The State of the Art of Preservice Teacher Education Programs and Supervision of Field Experience." *Action in Teacher Education* 2 (1980): 67–69.

National Council of Teachers of Mathematics. *Curriculum and Evaluation Standards for School Mathematics*. Reston, Va.: The Council, 1989.

———. *Professional Standards for Teaching Mathematics*. Reston, Va.: The Council, 1991.

Peterson, Penny L., Elizabeth Fennema, Thomas P. Carpenter, and Megan Loef. "Teachers' Pedagogical Content Beliefs in Mathematics." *Cognition and Instruction* 6 (1989): 1–40.

Shapiro, Phyllis P., and Agnes T. Sheehan. "The Supervision of Student Teachers: A New Diagnostic Tool." *Journal of Teacher Education* 37 (1986): 35–39.

Task Force on the Preparation of Teachers. *The Education of North Carolina's Teachers.* A Report to the 1986 North Carolina General Assembly. Raleigh, N.C.: The Board of Governors of the University of North Carolina, 1986.

Vacc, Nancy N., and Dorothy Russell. "Summer Student Teaching: Evaluation of a Pilot Program." *Action in Teacher Education* 8 (1992): 24–30.

12

Constructing Meaningful Understanding of Mathematics Content

Ruhama Even
Glenda Lappan

To TEACH the arithmetic-driven curriculum of the past, one needed little more than computational skill with the standard algorithms and a textbook to provide practice. That is no longer so. To prepare a teacher dedicated to helping children think mathematically requires a very different experience with mathematics from the traditional college course for elementary school teachers. With this in mind, a series of three innovative mathematics courses has been developed for undergraduate education majors in Michigan State University's Academic Learning Program, an alternative teacher education program designed for highly motivated prospective teachers. The program emphasizes the development of a thorough understanding of the subjects to be taught as well as a knowledge of how students learn in each subject and how to teach each subject effectively. Each Academic Learning student has a unique field experience that involves working with a mentor teacher and a classroom of children each term (including student teaching) for two years. Perry Lanier directs the project. Glenda Lappan is associate director and principal designer and instructor of the sequence of mathematics courses. Pamela Schram and Sandra Wilcox are the project's researchers. Ruhama Even participated in the conceptualization and development of the sequence of mathematics courses and the research instruments. Although the mathematics content of the courses emphasizes the goal of integrating and connecting mathematics topics, each course highlights one of three different areas of mathematics: number theory, geometry, or probability and statistics.

This work is sponsored in part by the National Center for Research on Teacher Education, College of Education, Michigan State University. The National Center for Research on Teacher Education is funded primarily by the Office of Educational Research and Improvement, United States Department of Education. The opinions expressed in this paper do not necessarily represent the position, policy or endorsement of the Office or the Department.

As we set out to design the mathematical component of the program, we identified the obstacles to change that we were likely to face—the beliefs and dispositions that the students bring with them: (1) elementary mathematics curricula driven by computational skill as the major goal, (2) mathematical knowledge as rule bound and unconnected, (3) teaching as telling and learning as memorizing. These beliefs and dispositions are not consistent with a modern set of goals for the study of mathematics nor with the needs of students. In order to challenge these beliefs and dispositions, the mathematics experience at the university has to cause our students to examine their fundamental beliefs about such questions as, What is mathematics? What does it mean to know mathematics? What mathematics do elementary school children need to study? How do children learn mathematics? What is the role of the teacher in the mathematics classroom?

In this paper, we describe the sequence of the innovative mathematics courses from the standpoint of *developers*—the guidelines we followed in making content decisions—and from the standpoint of *teachers*—the environment we created to help our students examine their deeply held beliefs in light of new experiences. This paper is organized around the joint themes of mathematics and teaching.

THE GOAL: GOOD MATHEMATICS—TAUGHT WELL

Preservice teachers' own experiences as learners furnish the data they use to make sense of what mathematics is and how it should be taught. Hence, the learning environment in the three-term sequence of classes had to be constructed in such a way that students experienced mathematics much as their own students might.

What do we mean by "good mathematics—taught well"? This phrase was first used in the Middle Grades Mathematics Project's final report to the NSF for grant number MDR8318218. When making decisions about which mathematical ideas to pursue in the courses, we asked ourselves many questions: Is this good mathematics? Is it important? What does knowing this idea enable a student to do? To what is it connected? How does it relate to the big mathematical ideas for elementary and middle school children? How does the content selected represent mathematics to the preservice teachers? Does the content require students to engage in doing mathematics? We were most concerned with the following three facets of mathematics: (1) mathematical content, (2) doing mathematics, and (3) mathematical connections. These facets are not independent of one another but rather are interrelated. Still, each one is important enough to be highlighted separately. Under mathematical content, we discuss the main topics and mathematical problem solving; under doing mathematics, we discuss abstraction, reasoning in mathematics, unique answers, and time spent on problems. The topic mathematical connections is discussed with a focus on representations and applications.

But "good mathematics" is not enough. Good mathematics has to be taught well. In planning the instruction for the series of mathematics courses, we were guided by three main principles: (1) the use of problem situations, (2) periodic reflections, and (3) an emphasis on the community. A detailed description follows of the main mathematical and instructional themes that guided us in the development of the courses.

Good Mathematics ...

Mathematical content

Mathematical knowledge includes understandings of particular topics, procedures, and concepts and the relationships among them. These understandings and relationships are what most people usually refer to when they talk about mathematical knowledge. Since this aspect of knowledge of mathematics is fairly familiar, we describe it briefly.

Main topics. We took an overall integrated approach to mathematics, but each term had a major emphasis that allowed us to probe ideas in depth. The three main themes represent important topics in the discipline of mathematics as well as in a desired elementary or middle school curriculum (e.g., National Council of Teachers of Mathematics 1989). The first term centered on the structure of number and number relationships; in the second term, the main theme was geometry; and in the last term, the emphasis shifted to data analysis, interpretation, and decision making. In each term, connections were made among number, geometry, and probability and statistics.

Mathematics and problem solving. Many "mathematics for elementary teachers" textbooks start with a chapter on problem solving. Then the following chapters concentrate on different topics (number sets, operations on numbers, geometry, probability and statistics, to name the most common). This approach seems to imply that problem solving and mathematics are two different activities: First you do problem solving and then you do mathematics. We wanted to send the message that problem solving and mathematics are not separate. Therefore, we started with a problem that was big enough to serve as a "problem solving" situation, but its solution was closely related to the mathematical topic to be taught—number theory. This problem—the Locker Problem—is discussed later in the paper. Throughout the three courses, we used the strategy of presenting to our students "big problems" that were related to the mathematics topic at hand. Even though the problems were related to the main topics being studied, they were "problems" in the sense that no direct, immediate, single way existed to solve them. So problem solving was integrated naturally into the courses.

Doing mathematics

Another part of mathematical knowledge includes understanding what it means to do mathematics. Many prospective elementary school teachers think that knowing mathematics means mastering a given set of facts, rules, and procedures (Ball 1988; Madsen-Nason 1988; Stodolsky 1987; Thompson 1984). To those who see mathematics this way, doing mathematics means recalling the appropriate facts, rules, or procedures. If the situation does not look familiar, one cannot use recall and thus feels unable to solve the problem. Conversely, if the situation looks familiar, recalling facts without understanding may lead to misuse of the recalled information. A belief that mathematics is more than mastering a given set of facts, rules, and procedures is not sufficient. Preservice teachers need to have ideas about how to structure classrooms so that they can help their students develop understanding. Since experience is a powerful teacher, it makes sense that these preservice teachers need to learn by experiencing mathematical ways of thinking, reasoning, analyzing, abstracting, generalizing, proving, and applying in environments that model good instruction.

Abstraction. Abstraction is a major component of doing mathematics. Mathematical concepts are abstract and "coming to know" in mathematics means, in many instances, abstracting from a variety of models and situations the important characteristics of a concept while ignoring the irrelevant ones. This approach guided us in our work. The students were often supplied concrete materials with which to work and were presented with various situations in which they encountered the same concept. The following example dealing with the concept of distance illustrates this point.

> Everybody knows what a distance between two points means. For example, given the following points on a grid (fig. 12.1), the distance between A and B is 4 units. The distance between A and C—$\sqrt{34}$—is a little harder to calculate; the Pythagorean theorem is needed. But suppose that the grid represents a map of city streets. You are in place A and need to get to place C. Now what's the distance between A and C? Is using the Pythagorean theorem appropriate in this case?

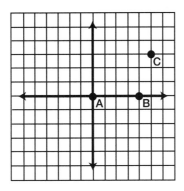

Fig. 12.1

Our students decided that "distance" in "taxicab geometry" should be defined to be the shortest path between two points. This definition is, of course, appropriate for distances in both Euclidean geometry and taxicab geometry, even though the distances may differ. In considering the same concept in two different geometries, one of which is unfamiliar, the students needed to abstract the meaning of distance, reaching a higher level of understanding of a concept they have taken for granted.

Reasoning in mathematics

Generalization—starting from specific cases to find a general rule—is central to doing mathematics. One way to look at algebra, for example, is as generalized arithmetic. But generalization is not limited to algebra. Whenever we deal with relationships and look for patterns, we deal with generalizations. The general rules can be described algebraically, geometrically, graphically, or verbally. Investigating a situation by checking specific cases is a powerful strategy. Many discoveries are made by inductive reasoning. Looking at specific cases helps in understanding a situation and in seeing why a conjectured rule should hold.

Looking for patterns and describing the general rules by using inductive reasoning was an important part of all the mathematics courses. From our experience, many prospective elementary school teachers try to solve problems by searching for the appropriate formula. Their beliefs about what mathematics is and about how one solves a mathematics problem and their conception of themselves in relation to mathematics shape this behavior. We wanted our students to experience inductive reasoning as a tool for solving problems in mathematics. Therefore, gathering data, checking specific examples, looking for patterns, and making conjectures based on generalizations were an important part of the courses. We can consider this approach as looking at mathematics as an empirical science in order fully to appreciate mathematics as a deductive science. The following example, called the Locker Problem, serves as an illustration (for a thorough discussion of this problem, see House [1980, p. 20]):

> In a certain high school there were 1000 students and 1000 lockers. Each year for homecoming the students lined up in alphabetical order and performed the following strange ritual: The first student opened every locker. The second student went to every second locker and closed it. The third student went to every third locker and changed it (i.e., if the locker was open, he closed it; if it was closed, he opened it). In a similar manner, the fourth, fifth, sixth, ... student changed every fourth, fifth, sixth, ... locker. After all 1000 students had passed by the lockers, which lockers were open?

The Locker Problem is really a problem. Students cannot solve it by recall, since the Locker Problem does not look like any familiar type of "story problem." One might guess and check—popular guesses are prime-number lockers, the first locker, and the last locker. It is easy to check that the first locker remains open, but how about the last one? Prime numbers don't seem to work (check 3 or 7, for example). It is clear that we have a *problem*. Someone in the class suggests that we see what happens with ten lockers. The class agrees that solving a simpler and more manageable problem might lead to some understanding of what is involved. Working in small groups, they "open" and "close" ten lockers: lockers 1, 4, and 9 remain open. Then they do the same with twenty lockers—1, 4, 9, and 16 are open. Sooner or later, each small group in the class has a conjecture: Either that all the open lockers are square numbers or that the differences between the open lockers are consecutive odd numbers.

Most prospective teachers are quite happy with their surprising solution and are willing to predict at this point which lockers are open. Since inductive reasoning is used in everyday life as a means for making predictions (Martin and Harel 1989), most students see this stage as the final solution of the problem. But can we really be sure that the pattern continues? Why? To make sure that it does, deductive reasoning should be used to construct a convincing supporting argument.

The questions mentioned above, in addition to some others that explore the relationship between a student's number and the numbers of the lockers visited, are assigned as homework. The next day, a whole-group discussion takes place. Many students have discovered that the relationship between the students' numbers and the numbers of the lockers they visited can be described as the relation between factors and multiples. Throughout the discussion, it becomes clearer that open lockers are the ones that have an odd number of factors. Do all square numbers have an odd number of factors? Why? Why do nonsquare numbers have an even number of factors? Investigating these questions by exploring factor pairs for some specific number (e.g., factor pairs for 24 are 1 and 24, 2 and 12, 3 and 8, 4 and 6. Factor pairs for 25 are 1 and 25, 5 and 5) clarifies why square numbers (and only square numbers) have an odd number of factors.

How about the other conjecture? Are the differences between open lockers consecutive odd numbers? Can we show that this conjecture is true? A pictorial representation (see fig. 12.2) can offer a convincing argument. The number 1 is represented by one dot at the upper left corner. By adding the number 3, which is represented by three dots, we can form the number 4—a 2 × 2 square. Then, by adding the number 5 (five dots), we can form the number 9—a 3 × 3 square. We can verify that this process continues for any given sum of consecutive odd numbers starting with 1. The result is always a square number.

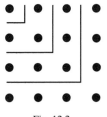

Fig. 12.2

Inductive reasoning, important as it is to mathematical activity, is not a sufficient explanation for the existence of a rule, nor is it a proof (unless we can check all cases—a strategy that is used more and more in modern mathematics with the power of new technology). In order to transform a conjecture into a theorem, we need to use mathematically appropriate and acceptable ways to construct either a logical verification or a counterexample. "Deductive reasoning is the method by which the truth of a mathematical assertion is finally established" (NCTM 1989, p. 143). But many prospective teachers do not see the need for deductive reasoning (Even 1989; Martin and Harel 1989). Providing a sound mathematical explanation was an important part of the courses. The questions Why? and How do you know that? were asked often. We were not after a formal proof that uses the "appropriate" format as is often so in high school Euclidean geometry, but rather we wanted to develop mathematical ways of thinking and reasoning at a more informal level. The observation that lockers with square numbers remain open when thirty lockers are checked does not prove that this result will always hold. Showing that square numbers have an odd number of factors and relating this fact to the problem does provide a convincing argument for the conjecture.

Unique answer. Common misconceptions among elementary school teachers are that every mathematics problem has one and only one answer and that there is only one way to get this answer. Not only are these misconceptions false representations of mathematics, but this way of thinking causes difficulties with the learning of mathematics. It encourages recall and memorization of *the* right way to solve problems instead of creativity and independent thinking.

We encouraged diverse approaches and views of a problem situation throughout the courses. For example, the students were presented with figure 12.3 on the overhead projector and were given the following problem:

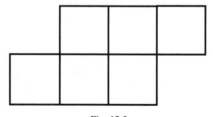

Fig. 12.3

Assume that the edge of the small squares is 1 unit in length. Add squares so that the figure has a perimeter of 18. When squares are added, they must meet exactly along at least one edge of the figure.

Surprising to many people, the fact that the perimeter is fixed does not imply that the shape of the solution figure is fixed nor that all solution figures have the same area. Experience with problems that have more than one solution raised questions such as, Are any solutions more interesting than the others? Do any solutions have special aspects, such as largest or smallest? If so, do these solutions have a special significance? Since problems that arise in the real world are often ill defined or have more than one interpretation or solution, these problems show an aspect of mathematics that is important but rarely experienced in traditional mathematics courses.

Time Spent on a Problem

A common belief about solving problems in mathematics is that if one cannot solve a problem in a very short time, one will not be able to solve that problem at all (e.g., Schoenfeld 1988). In order to change this belief about mathematics, we gave the prospective teachers many opportunities to spend much more time than usual in solving a single problem. We did it in two different but complementary ways. One way was to spend several class periods on the same problem. But having intending teachers experience different mathematics in class is not sufficient. Accustomed to give up on problems very quickly, they must be encouraged to change their behavior outside class as well. So to help our students in another way, we occasionally chose an important or interesting problem from their homework assignment and asked about its solution in class. We were not after the final answer. Rather, we wanted people to discuss their attempts, findings, and difficulties in order to help them make some progress toward a solution. Spending time discussing work that had been done on the problem without arriving at a solution, or even evaluating students' attempts to solve the problem, made it clear to the students that giving up after a short trial was not acceptable in these courses and that they were responsible for solving the problem. The latter approach also implies that the students can do the problem and therefore should try. The Infinite Forest Problem illustrates this idea. The problem was posed as follows:

> Suppose that you have an infinite geoboard and that on each of the lattice points, except the one at the origin, there is a tree with a trunk that is only as wide as a line. You are standing on the origin. Is there a straight-line path that you can take from the origin that will allow you to walk forever in the forest and not hit a tree?

After we posed this problem, it was discussed by the class for parts of several meetings before one student put forward an idea that stimulated the class to consider the implication of hitting a tree while walking. From this point on, the solution was easy for the class. They said that hitting a tree implied that the path hit another lattice point. This meant that that path had a rational slope. They then constructed a length equal to the square root of 2 perpendicular to the x-axis at the point $(1, 0)$. This construction gives a path that has an irrational slope, which implies that it cannot hit another lattice point.

Mathematical connections

Another characteristic of mathematical knowledge is rich connections (Even 1990; Hiebert and Lefevre 1986). One cannot understand a mathematical concept in isolation. Connections to other concepts, procedures, and pieces of information deepen and broaden knowledge. Two important aspects of connections that we emphasized in our courses were the use of different representations and applications both within mathematics and between mathematics and other areas of study.

Representations. Representing ideas and problems in different ways—geometrically, verbally, numerically, algebraically, or physically—allowed the students to see how different representations give different insights into problem situations (Dufour-Janvier, Bednarz, and Belanger 1987; Lesh, Post, and Behr 1987). Developing flexibility in representing ideas in different ways and interpreting among different representations were for us important parts of developing mathematical power. The continued work on the Perimeter of 18 Problem illustrates multiple representations and their power.

After sharing and discussing the different solutions the class found, related questions arose: What is the fewest number of squares that must be added to make the perimeter 18? What is the greatest number of squares that you can add and keep the perimeter 18?

A close analysis of what happens to the perimeter when one square is added to a figure shows that if only one edge of the square touches one edge of the figure, the perimeter grows by exactly two units. If the square is added in a "corner" and two edges touch two edges of the figure, the perimeter does not change (although the area does). When three edges of the added square touch three edges of the figure, the perimeter is two units smaller. This information makes clear that the shape of the resulting figure with the most squares should be a rectangle. But which one? Using tiles, the students construct the following rectangles, all with a perimeter of 18: 1×8, 2×7, 3×6, 4×5. They check and find out that the 4×5 rectangle has the greatest area—20 square units.

Is this the answer to the problem? Well, it depends on the domain in which we are working. For the given plastic tiles, the 4×5 rectangle is the figure with the most squares (greatest area) that still has a perimeter of 18. But what if we allow the dimensions of the rectangle to be any real number? Further investigation of the four rectangles with perimeter 18 that can be made from the square tiles shows that as the bottom edge and the side edge of the rectangles become closer in length, the area grows. This observation leads students to conjecture that the solution figure is a square. What is the length of the square's side? Some suggestions from students were 4.5 and $\sqrt{20}$. But most students were not sure.

Graphing the area versus the length of each of the rectangles (fig. 12.4) suggests an answer. The graph seems symmetric and suggests that the maximum area is midway between 4 and 5—4.5. A rectangle with perimeter 18 and length 4.5 is, of course, a square with area $4.5^2 = 20.25$. This answer seems reasonable, but can we really be sure that the maximum area is obtained at 4.5? Maybe between 4 and 5 the graph goes down. Maybe it just seems to be a parabola but it is actually not.

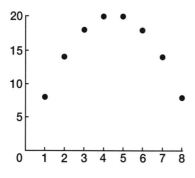

Fig. 12.4

An algebraic representation can, without using calculus, give a definite answer to this dilemma about where the maximum occurs. Compare the area of the square with perimeter 18 to the area of any rectangle with perimeter 18 (fig. 12.5). Let's call the width of the rectangle x, then the length is 9 minus x. The square has side 4.5. The square is composed of parts A and B; the rectangle of parts B and C. Since part B is common to both, we need to show that the area of part A is greater than the area of part C.

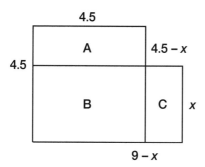

Fig. 12.5

Symbolically, we need to show that 4.5(4.5 minus x) > x(9 minus x minus 4.5). This inequality can be written as 4.5(4.5 minus x) > x(4.5 minus x). Since x < 4.5 (the width of the rectangle is shorter than the side of the square), the inequality holds. Therefore, among all rectangles with perimeter 18, the square has the largest area.

The foregoing is an example of a problem in which moving from one representation to another contributed to a construction of richer and deeper knowledge about perimeter, area, and the relations between them.

Applications. One characteristic of problem solving is application. We thought of applications as problems that require mathematical thinking in their solution and that come from a real-world situation. Such problems may call for problem solving that is as creative and as challenging as those that wear the label "problem-solving problem." The distinction for us is the requirement of a realistic context.

We paid special attention to applications, since having to apply existing knowledge in a new situation, whether inside or outside mathematics, sheds a new light on old knowledge and creates new connections and relationships among different pieces of knowledge. We posed many problems for which students needed to integrate and apply their knowledge. For example, in the last course of the sequence the students learned new ways of looking at and interpreting data. Then they were asked to discuss and agree on a related set of questions that the data could help answer. The class designed a questionnaire to gather the data, planned and carried out the data gathering, analyzed the data, and organized the data for a presentation of what the information said about their original problem. The class decided that they wanted to know something about the typical MSU female and male student. They stated their questions thus: Who are you, Mr. MSU? Who are you, Ms. MSU?

...Taught Well

Good mathematics is a necessary component of a desired mathematics course. But it is not sufficient. Good mathematics should be taught well. Simply telling students that all square numbers have an odd number of factors and therefore the lockers with square numbers are the ones that remain open at the end wouldn't have the same learning effect as the experience we described earlier. Some of the principles that guided us were implicitly described when we talked about the mathematical aspect of the sequence. Next we would like to discuss the three main aspects that characterized the instruction in these courses: (1) use of problem situations, (2) periodic reflections, and (3) an emphasis on community.

Problem situations

In the traditional mathematics curriculum, mathematical facts and procedures are often studied until mastered and then applied to a specified set of problem types. The organization of textbooks frequently gives the learner clues that reduce problem solving to matching a pattern in a given example. The results of this kind of mathematics education are that all too often students have computational skills but no idea when to use these skills or what the results mean in a given context. The results of the 1983 National Assessment of Educational Progress (Carpenter, Lindquist, Matthews, and Silver 1983) illustrate this problem well.

In our courses we took as a primary goal to embed the mathematics in situations or contexts that help give meaning to the resulting concepts, rules, or procedures. Research in learning supports the notion that humans process information and are more likely to be able to recall and use this information if it is contextualized. Brown, Collins, and Duguid (1989, p. 32), for example, argue that

> the activity in which knowledge is developed and deployed ... is not separable from or ancillary to learning and cognition. Nor is it neutral. Rather, it is an integral part of what is learned. Situations might be said to co-produce knowledge through activity. Learning and cognition ... are fundamentally situated.

The examples given in this paper—the Locker Problem and the Perimeter of 18 Problem—illustrate this point. These settings became an often-used way of referring to the mathematical ideas embedded in the problem. Another result of this approach was the complete absence of comments like, "Why do we need to know this?" or "What's this good for?" Students develop a different notion of what mathematics is about if they are constantly confronted with situations from which mathematics arises rather than being given the record of other peoples' rules and algorithms in an abstract form (Dewey 1904). NCTM's *Curriculum and Evaluation Standards* (1989) has two overall goals for students: (1) to learn to value mathematics, and (2) to become confident of their ability to do mathematics. Situated mathematics can contribute to each of these goals by presenting the students with interesting and meaningful mathematics that is open to various and different levels of solutions.

Community

Although communication of mathematical ideas is an important part of the experiences that all students should have in mathematics classrooms (Lappan and Schram 1989), communication seemed to us to be absolutely necessary for intending teachers. Our students needed to learn mathematics, but they also needed to become sensitive to the role of communication in clarifying their thoughts and in expanding their repertoire of ways of thinking. In addition, we were concerned that our students learn to listen to others and try to make sense of their ideas. These concerns led us to structure our classroom as a community of learners with considerable responsibility for judging, validating, and helping others. The teacher did not give the final verification that the ideas put forward were the "correct" ones. Verification was the responsibility of the whole group. The teacher's role was to pose interesting mathematical tasks for the students to consider individually, in small groups, and as a whole class. She also asked questions that helped the class to learn to value convincing arguments and to demand that mathematics make sense.

Creating a classroom that was a community was not a simple task. Our students' history dictated to them what a mathematics class should look like. Moving them to a new, desired, active role was difficult. It was the long-term involvement with these students that allowed us to move them from frustration at wanting "the answer" to not only accepting a different role and different demands but also being unwilling to "take the teacher's word." The students changed during the time from the first term to the third one. They took responsibility for solving problems and did not consider the teacher as the authority on correct answers. Asking the teacher to verify answers or give the right answers, which was common in the first term, became rare during the last term. These students insisted on understanding—on making sense of the mathematics. They had learned to value small-group work, individual effort, and the power of the community to decide what should be accepted as valid in our growing repertoire of mathematical knowledge.

Periodic reflections

The integrating of reflection was a regular feature in the mathematics courses. Questions were asked on a daily basis that focused the students' attention on connections to what had been studied and to what was coming up, but we also took the time periodically to reflect as a group on the mathematics being studied. This reflection took the form of creating concept maps of domains of knowledge, generating lists of the current working "theorems" and the conjectures that still remained to be supported or refuted, developing different forms of representation of ideas and concepts that we had studied, and considering what each representation did to help us understand or explain a problem. To summarize, these reflection periods were intended to support the development of a well-integrated, connected knowledge of mathematics. They were generative. The goal was not to produce a list of facts or procedures studied; it was to find new ways to organize and conceptualize the mathematical experiences that the class had shared.

CONCLUSION

The purpose of school includes preparing young people for full participation in the society and culture of the modern world. Examining the role of mathematical thinking in our society forces us to reconsider the existing goals of the K–12 mathematics curriculum. In particular, the goals of the elementary school mathematics curriculum that center on developing computational proficiency with paper-and-pencil algorithms must change. Mathematics is a dynamic cultural invention that grows and changes as the needs and interests of society evolve. In the modern world, this evolution of mathematical knowledge and society's dependence on mathematical ideas have become a revolution. A veritable explosion of mathematical thought and invention has been spurred by the invention of computing devices that make possible approaches to mathematics that were unthought of in the past. This change in mathematics has mirrored a change in our society and culture that makes the mathematical currency of the modern world the skill and disposition to see the world mathematically—to create mathematical models of problem situations, to manipulate these models, and to interpret the results in relation to the original problem.

How can we achieve this change in school mathematics? There is, of course, no simple answer to this question, nor is it a new issue. The following excerpt, taken from Dienes's (1960, p. 1) book *Building Up Mathematics*, is as relevant now as it was thirty years ago:

> Let us face it: the majority of children never succeed in understanding the real meanings of mathematical concepts. At best they become deft technicians in the art of manipulating complicated sets of symbols, at worst they are baffled by the impossible situations into which the present mathematical requirements in schools tend to place them.

From the beginning of the century, Moore (1926), Brownell (1935, 1947), Van Engen (1953), Bruner (1960, 1961), Dienes (1960), Biggs (Biggs and MacLean 1969), Skemp (1987), and others called for change in school mathematics. Still, the way mathematics is taught at school has changed little.

Change in teaching and learning depends heavily on the teacher. Conversely, teaching, by its very nature, includes fundamental barriers to change (Cohen 1988). Teaching, Cohen says, is a practice of human improvement wherein one human being tries to improve the ideas, capacities, emotional states, or organization of others. Practices of human improvement are hard to manage. Therefore, most practitioners and clients tend toward conservative strategies. Most practitioners of human improvement, such as therapists and organization consultants, have some protection from the social constitution of their work; they can choose their clients and are not expected to succeed without the cooperation of their clients. But teachers face the internal problems of teaching with little or no such protection. Therefore, the tendency toward conservative approaches to practice is even stronger. Cohen concludes that the internal problems of teaching that result from teaching's being a practice of human improvement, compounded by such external conditions as finance and organizations, cause great difficulties for teachers in implementing instructional reforms.

Although substantial change in teaching is difficult to achieve, some change lends itself to implementation. "Effective teaching" strategies are an example. These strategies are aimed at "methodological refinement" (Aronowitz and Giroux 1985), they provide an algorithmic approach to improving instruction and are therefore easy to adopt, and they do not challenge the existing structure of authority in classrooms and schools. Still, using effective-teaching strategies may change teachers' behavior in some way but by itself does not cause teachers to teach for understanding, nor does it make students' learning meaningful.

Past unsuccessful reforms and the internal and external barriers to change in mathematics curriculum and teaching show us that substantial change in school mathematics cannot occur easily. A good preservice education for teachers is a necessary (although not sufficient) aspect of learning to teach in ways that will enable teachers to create new, desirable learning environments for students. To be effective, the mathematical experiences must cause preservice teachers to build powerful mathematical schemata and to examine their deeply held beliefs about mathematics as a discipline, how it is learned, and the role of the teacher.

Recognizing that change in beliefs and practices is difficult to effect, we have worked from the premise that teachers need what we want for students. If students in elementary and middle school are to learn in environments that support the development of mathematical power as described in the *Curriculum and Evaluation Standards*, teachers themselves need to know mathematics and experience learning in ways that build a deep and flexible understanding of what mathematics is and what it means to do mathematics. To accomplish this goal, we developed the three mathematics courses described. This paper is intended to give a picture of the development process and of the kinds of experiences that the courses offered our students. We hope that our experience will be of benefit to others who engage in research and the development of preservice teachers of mathematics.

REFERENCES

Aronowitz, Stanley, and Henry A. Giroux. *Education under Siege: The Conservative, Liberal, and Radical Debate over Schooling.* South Hadley, Mass.: Bergin & Garvey Publishers, 1985.

Ball, Deborah L. "Knowledge and Reasoning in Mathematical Pedagogy: Examining What Prospective Teachers Bring to Teacher Education." Doctoral dissertation, Michigan State University, 1988.

Biggs, Edith E., and James R. MacLean. *Freedom to Learn: An Active Learning Approach to Mathematics.* Don Mills, Ont.: Addison-Wesley, 1969.

Brown, John S., Allan Collins, and Paul Duguid. "Situated Cognition and the Culture of Learning." *Educational Researcher* 18 (January-February 1989): 32–42.

Brownell, William A. "The Place of Meaning in the Teaching of Arithmetic." *Elementary School Journal* 47 (January 1947): 256–65.

———. "Psychological Considerations in the Learning and the Teaching of Arithmetic." In *The Teaching of Arithmetic,* Tenth Yearbook of the National Council of Teachers of Mathematics, edited by W. D. Reeves, pp. 1–31. New York: Columbia University, Teachers College, Bureau of Publications, 1935.

Bruner, Jerome S. "The Act of Discovery." *Harvard Educational Review* 31 (Winter 1961): 21–32.

———. "On Learning Mathematics." *Mathematics Teacher* 53 (December 1960): 610–19.

Carpenter, Thomas P., Mary M. Lindquist, Westina Matthews, and Edward A. Silver. "Results of the Third NAEP Mathematics Assessment: Secondary School." *Mathematics Teacher* 76 (December 1983): 652–59.

Cohen, David K. *Teaching Practice: Plus Ça Change.* Research report no. 88-3. East Lansing, Mich.: Michigan State University, National Center for Research on Teacher Education, 1988.

Dewey, John. "The Relation of Theory to Practice in Education." In *John Dewey on Education,* edited by R. Archambault, pp. 313–38. Chicago: University of Chicago Press, 1904.

Dienes, Zoltan P. *Building Up Mathematics.* London: Hutchinson Education, 1960.

Dufour-Janvier, Bernadette, Nadine Bednarz, and Maurice Belanger. "Pedagogical Considerations Concerning the Problem of Representation." In *Problems of Representation in the Teaching and Learning of Mathematics,* edited by C. Janvier, pp. 109–22. Hillsdale, N.J.: Lawrence Erlbaum Associates, 1987.

Even, Ruhama. "Prospective Secondary Teachers' Knowledge and Understanding about Mathematical Functions." Doctoral dissertation, Michigan State University, 1989.

———. "Subject Matter Knowledge for Teaching and the Case of Functions." *Educational Studies in Mathematics* 21(December 1990): 521–44.

Hiebert, James, and Patricia Lefevre. "Conceptual and Procedural Knowledge in Mathematics: An Introductory Analysis." In *Conceptual and Procedural Knowledge: The Case of Mathematics,* edited by James Hiebert, pp. 1–27. Hillsdale, N.J.: Lawrence Erlbaum Associates, 1986.

House, Peggy A. "Making a Problem of Junior High School Mathematics." *Arithmetic Teacher* 28 (October 1980): 20–23.

Lappan, Glenda, and Pamela W. Schram. "Communication and Reasoning: Critical Dimensions of Sense Making in Mathematics." In *New Directions for Elementary School Mathematics,* 1989 Yearbook of the National Council of Teachers of Mathematics, edited by Paul R. Trafton, pp. 14–30. Reston, Va.: The Council, 1989.

Lesh, Richard, Thomas Post, and Merlyn J. Behr. "Representations and Translations among Representations in Mathematics Learning and Problem Solving." In *Problems of Representation in the Teaching and Learning of Mathematics,* edited by C. Janvier, pp. 30–40. Hillsdale, N.J.: Lawrence Erlbaum Associates, 1987.

Madsen-Nason, Anne L. "A Teacher's Changing Thoughts and Practices in Ninth Grade General Mathematics Classes: A Case Study." Doctoral dissertation, Michigan State University, 1988.

Martin, W. Gary, and Guershon Harel. "Proof Frames of Preservice Elementary Teachers." *Journal for Research in Mathematics Education* 20 (January 1989): 41–51.

Moore, Eliakim Hastings. "On the Foundations of Mathematics." In *A General Survey of Progress in the Last Twenty-five Years,* First Yearbook of the National Council of Teachers Mathematics, edited by Raleigh Schorling, pp. 32–57. n.p.: The Council, 1926.

National Council of Teachers of Mathematics. *Curriculum and Evaluation Standards for School Mathematics.* Reston, Va.: The Council, 1989.

Skemp, Richard R. "Relational Understanding and Instrumental Understanding." *Arithmetic Teacher* 26 (November 1987): 9–15.

Stodolsky, Susan. "The Subject Matters, Instruction in Math and Social Studies." *Notes and News.* [Michigan State University, Institute for Research on Teaching], (29 May 1987): pp. 2–3.

Thompson, Alba G. "The Relationship of Teachers' Conceptions of Mathematics and Mathematics Teaching to Instructional Practice." *Educational Studies in Mathematics* 15 (May 1984): 105–27.

Van Engen, Henry. "The Formation of Concepts." In *The Learning of Mathematics: Its Theory and Practice,* Twenty-first Yearbook of the National Council of Teachers of Mathematics, edited by Howard F. Fehr, pp. 69–98. Washington, D.C.: The Council, 1953.

13

A Model Preservice Program for the Preparation of Mathematics Specialists in the Elementary School

Chuck McNerney

In 1983, the Office of Educational Research and Improvement of the U.S. Department of Education, the National Council of Teachers of Mathematics, and the Wisconsin Center for Educational Research sponsored a conference on school mathematics that was held in Madison, Wisconsin. Among the recommendations in the chair's report of this conference is the following (U.S. Department of Education 1984, p. 17):

> In elementary school, specialist teachers of mathematics should teach all mathematics beginning no later than Grade 4 and supervise mathematics instruction at earlier grade levels.

In 1989, the National Research Council (1989, p. 64) continued to encourage the use of mathematics specialists in the elementary schools:

> The United States is one of the few countries in the world that continues to pretend—despite substantial evidence to the contrary—that elementary school teachers are able to teach all subjects equally well. It is time that we identify a cadre of teachers with special interests in mathematics who would be well prepared to teach young children both mathematics and science in an integrated, discovery-based environment.

Other references to the need for mathematics leader-specialists can be found in Mathematical Association of America (MAA) (1983), Dossey (1984, pp. 3, 50), MAA (1988), MAA (1991), and Bruni (1991, p. 7).

A PROBLEM WITH THE PREPARATION OF MATHEMATICS SPECIALISTS

It appears that with the exception of the 1983 Wisconsin group that recommended the development of model preservice programs for persons seeking to be elementary mathematics specialists (U.S. Department of Education 1984, p. 20), most of the recommendations mentioned above are directed toward in-service, rather than preservice, education. This emphasis creates a problem in that a considerable amount of in-service work is necessary to teach generalists adequate mathematics to enable them to become mathematics specialists. Contributing to this problem is that many, if not most, of the preservice programs in this country do not provide adequate preservice mathematics preparation of prospective elementary school teachers to teach the content currently being recommended by the National Council of Teachers of Mathematics's *Curriculum and Evaluation Standards for School Mathematics* (1989). As of June 1987, fourteen states had *no* course requirements in mathematics for the certification of elementary school teachers. In an additional ten states, the mathematics requirement for elementary school certification was determined by an approved competency-based program or by degree-granting institutions (Council of Chief State School Officers 1990). Prospective elementary school teachers in the other states had for the most part taken at most two mathematics courses and a single mathematics methods course as part of their undergraduate certification program. The mathematics courses, with the exception of those required for general education, generally fall under the umbrella of what was considered to be Level I CUPM (Committee on Undergraduate Programs in Mathematics) courses for the preparation of elementary school teachers (MAA 1983). It is extremely difficult to build on these courses in an in-service setting to prepare the generalist to become a mathematics specialist, especially if the teacher has been away from mathematics for any significant length of time.

A MODEL PRESERVICE PROGRAM

I would like to suggest to the mathematics education community that the profession is capable of preparing mathematics leaders for the elementary schools while preparing the prospective elementary teacher to enter the profession. This suggestion is based on my involvement with a program that has been in place in Colorado for the past four years.

Several forces led to the development of this program. In 1985, the Colorado state legislature, following the precedent of states such as Michigan, which does not sanction the offering of majors in education except in special cases (Johnston et al. 1989, p. 141), passed legislation that, among other things, eliminated all undergraduate education majors in Colorado. Incoming students who wished to become elementary or middle school teachers in Colorado were required to pursue a major in any one of several designated academic areas that include the liberal arts. Each university involved in teacher education at the undergraduate level was then expected to prepare appropriate academic undergraduate programs for prospective elementary and middle school teachers.

The Department of Mathematical Sciences at the University of Northern Colorado, following the state mandate, designed a baccalaureate program in mathematics with an emphasis in elementary or middle school education that consisted of thirty-eight semester hours of mathematics, statistics, and computer science. Twelve of these hours are professionalized content courses. The first three courses in this sequence, Fundamentals of Mathematics I and II and Informal Geometry, are built around the COMET (Committee on the Mathematical Education of Teachers) Level 1 recommendations. Each course heavily emphasizes the use of manipulatives, calculators, problem solving, and cooperative learning in the study of mathematics. The second course also includes the study of LogoWriter and mathematics and science connections. This course is currently taught in a departmental model mathematics classroom with sixteen networked Centris 610 computers, a Barco projection system, and trapezoidal tables that can easily be moved together for doing experiments and using manipulatives. The third course is a study of informal geometry, which includes the use of Miras as a manipulative in the study of transformational geometry.

The last course in the professionalized content core is a study of topics in such areas as abstract algebra, set theory, logic, number theory, matrix algebra, computational mathematics, and others with an emphasis on mathematical connections. Mathematical structure is demonstrated and emphasized throughout the course in such areas as number theory (the ring of integers modulo n), logic (Boolean rings), matrix algebra (the ring of 2×2 matrices), set theory (the ring of all subsets of a given set with the appropriate operations), and finite group and field systems. Computer software is used to analyze finite structures for various group, ring, and field properties.

The remaining courses of the major are mathematics, statistics, and computer science courses such as calculus, linear algebra, discrete mathematics, the history of mathematics, introductory statistics, Pascal, and college geometry. These courses are blocked with the professionalized content courses so that the student can meet prerequisites in each track early in the program without carrying an inordinate number of difficult mathematics courses. For example, Fundamentals of Mathematics I and Calculus I are taken in the first semester of the students' freshman year and Fundamentals of Mathematics II and Calculus II are taken in the second semester of the program. Informal Geometry is then blocked with Linear Algebra or Discrete Mathematics, and so on.

The program is being modified to expand the major requirements to include a capstone mathematics education course to be taken during the students' senior year. The content in this course is built around the NCTM's *Professional Standards for Teaching Mathematics* (1991) and the NCTM's recommendations in its position paper *Mathematics Leaders in Elementary/Middle Schools* (1984).

The first students to complete the program graduated in May 1992. They participated in a general certification program, common to all prospective elementary or middle school teachers (grades K–5 or 6–8) at the university. In short, they will enter the profession as generalists with a content specialty in mathematics. The following courses constitute the mathematics program:

Pascal Programming	3 hours
Calculus and Analytic Geometry I	4 hours
Calculus and Analytic Geometry II	4 hours
Fundamentals of Mathematics I	3 hours
Fundamentals of Mathematics II	3 hours
Linear Algebra	3 hours
Discrete Mathematics	3 hours
Informal Geometry	3 hours
Introduction to Modern Geometry	3 hours
Topics in Mathematics for Teachers	3 hours
History of Mathematics	2 hours
Introduction to Statistical Analysis	3 hours
TOTAL	38 semester hours

Observations about This Program

The following aspects of the program merit further elaboration:

1. The primary emphasis of the mathematics content in this major is not to give prospective elementary and middle school teachers the mathematics content necessary for them to do their job in the elementary or middle school classrooms after graduation; this objective is secondary. Such preparation can be accomplished by designing a nine-to-fifteen-hour mathematics sequence in accord with the recommendations in the *Professional Standards for Teaching Mathematics* (NCTM 1991, pp. 135–39). These recommendations are designed for those at the elementary or middle school levels who will teach the content recommended for these levels in the *Curriculum and Evaluation Standards for School Mathematics* (NCTM 1989).

The primary objective of this model program is to provide a meaningful experience for students who are genuinely interested in mathematics and who also want to be elementary or middle school teachers. This program is not a watered-down experience in mathematics! Although the program participants are not required to take the same variety of mathematics courses as the other mathematics majors at our university, the two calculus courses, the discrete mathematics course, the linear algebra course, and the other courses common to all mathematics majors are taken with prospective secondary school mathematics teachers and liberal arts students. What distinguishes this major from the other majors is a professional component that emphasizes content suitable for the elementary and middle school levels and that models good pedagogy.

2. The overwhelming majority of the students in this program are female. The placement of these students in the elementary classrooms throughout this nation is invaluable. Imagine the message given to the young female child who is being taught by a female graduate of this program. We, as a nation, are struggling to overcome the ridiculous myth that women can't do mathematics. What better way to dispel this myth than to have female role models in the elementary classroom who know and enjoy mathematics and who are capable of stimulating both male and female elementary school students in the study of mathematics?

3. The program is viable and very much alive. Nearly one hundred students have chosen this program, and a suitable number of new candidates enter the program each year. Young men and women throughout this nation are genuinely interested in, and capable of doing, mathematics and want to teach in the elementary and middle schools. These young men and women come to the university looking for a program that will afford them an opportunity to study mathematics while preparing to enter the profession of teaching. In the past, this avenue of mathematical study was not open for most elementary and middle school teachers because they were required to major in education or to pursue a certification program in secondary school mathematics.

4. The students we are seeing in this program are for the most part well-qualified in mathematics. Most of the students have studied four years of high school mathematics; some have taken AP Calculus. High schools of all kinds throughout Colorado and other states have done a very good job of preparing these people to study mathematics at the college level. Some students are required to take remedial courses before they enter their degree program, but this is the exception, not the rule.

The students who are pursuing this degree are also doing quite well academically, possibly because they are studying mathematics for the right reason, namely, that they enjoy and are capable of studying mathematics. They are fully aware that they will not be teaching calculus, linear algebra, or discrete mathematics to their elementary school students. They also know that they will be certified as generalists who are expected to assume leadership roles in mathematics in the public schools after graduation, and this prospect excites them.

Benefits of the Program

Several benefits result from this program:

1. Undoubtedly, the greatest benefit is the provision of mathematics specialists for the elementary school classroom. I stated earlier that a considerable amount of in-service work is necessary to teach generalists adequate mathematics to enable them to become mathematics specialists. The graduate of this model program has a degree in mathematics that satisfies many, if not most, of the course requirements in the master's program for elementary school teachers put forth by COMET (MAA 1988). The COMET program includes courses in problem solving, calculators and computers in elementary school mathematics, concepts of geometry, and probability and statistics in the elementary school curriculum. The mathematics requirements of our model program more than satisfy the COMET requirements (and the mathematics requirements for the proposed COMET middle school degree). The emphasis on modeling good pedagogy in the present program and the addition of the new mathematics education course requirement will satisfy many of the requirements in the four mathematics education courses described by COMET. The point to be made here is

that this preparation takes place during the undergraduate degree program, not as part of a graduate program.

2. In addition, this program affords graduates the opportunity for employment outside the teaching profession should they decide to leave the profession and seek other employment. Some graduates of the program may aspire to teach in the middle school or the secondary school; these graduates can return to the university and take the courses required for such certification. Secondary school certification at our university involves taking another sixteen hours of mathematics, a secondary school methods course, and a student-teaching experience at the secondary school level.

3. The program also provides for research and development opportunities for both university staff and graduate mathematics education students. Innovation in both content and pedagogy is necessary in the professionalized content core. Presently, mentors, who are certified elementary and middle school teachers, are assisting the university instructors in the planning and teaching of the professionalized content courses. The content of the topics course is presently being developed, written, and piloted by two department members. Each of these activities and the follow-up studies of the program's graduates after they enter the profession are significant areas for research.

Criticisms of the Program

Few innovative programs go without their share of criticism. Some of the criticisms of this program are listed below:

1. It is difficult to graduate from this program in four years. Some students require more than eight semesters to complete their degree. A student can graduate with a bachelor of arts in mathematics in eight semesters if the student carries the equivalent of seventeen hours per semester. This is a considerable number of hours; hence, some students attend an on-campus summer session or take courses off campus to lighten their on-campus load during the academic year.

2. The major in mathematics is difficult. Some students have difficulty maintaining an above-average grade-point average in the mathematics courses. This problem is significant because the College of Education requires an overall grade-point average of 3.0 (on a scale of 0 to 4) for entry into professional teacher education.

3. Some critics maintain that elementary school teachers, being generalists by certification, should be broadly prepared as education majors and should not be required to pursue a liberal arts major. This argument is moot in Colorado, since public universities must follow the mandates of the state legislature. For a more complete and interesting discussion of this issue, see Johnston and associates (1989).

Program Development Considerations

As with most new programs, critical elements must be addressed during both the development and the implementation of the program. A partial listing of these considerations follows:

1. Colleges and universities must undertake the recruitment of minorities and underrepresented groups for this major. The program addresses the problem of the placement into the elementary school classroom of females who are mathematically capable, but a concerted effort must be made to recruit and retain other underrepresented groups.

2. Public school administrators must recognize the value of a content specialist in the elementary school and employ them accordingly. Public school administrators might argue that the content specialist cannot teach well in all areas of the elementary school curriculum. The generalist preparation of the content specialist in this program does not support this contention. Prospective elementary school teachers in mathematics programs should not only be recognized for having completed a difficult content degree but also be rewarded with jobs in a difficult job market.

3. A state or certification requirement of a liberal arts major or its equivalent for all prospective elementary school teachers is of tantamount importance. The bachelor's degree that emphasizes preparation for teaching elementary school as well as an academic area usually does not afford the in-depth exposure needed by the mathematics specialist. In many instances, the latter offers a superficial treatment of mathematics that is inappropriate for specialist preparation.

4. Cooperation is necessary between the College of Education and the College of Arts and Science. Students should have both a major advisor who is concerned with their mathematics and general education and a certification advisor who counsels them in the area of certification requirements. Both areas of advisement are important. Students need to begin their mathematics major and certification program during the first semester on campus, and they should be directed, at that time, to the appropriate courses.

5. The choice of courses for the major is significant. A traditional liberal arts emphasis or a secondary school emphasis in mathematics does not allow for the professionalized content core that elementary school teachers must have for both professional and certification purposes. This requirement can most easily be accomplished by the provision of a professionalized content core of courses that are required in the major.

6. The presence of mathematics educators with a specialty or interest in elementary teacher education in the department sponsoring the major is valuable in establishing both the content program and the rapport with the advisors in the College of Education. In some instances, the professionalized content core of twelve hours is best taught by mathematics educators.

CONCLUSION

If a program such as ours is to continue to provide specialists to the profession, we must convey a message to the school districts from which these students have graduated. Teachers must be told that they have done a good job in preparing these students for university mathematics work and should be encouraged to continue to do so. In addition, parents, teachers, administrators, and counselors throughout the system must be made aware that young men and women who are interested in mathematics can, and should be encouraged to, pursue a study of mathematics at the university level while preparing for elementary or middle school certification. Finally, principals and other administrators must be willing to hire the graduates of this program. Administrators must be convinced that these graduates who are generalists with a specialty in mathematics can be valuable additions to their staff.

REFERENCES

Bruni, James V. "One Point of View: We Must Have 'Designated Math Leaders' in the Elementary School!" *Arithmetic Teacher* 39 (September 1991): 7–9.

Council of Chief State School Officers. *State by State Indicators of Science and Mathematics Teachers.* Washington, D.C.: State Education Assessment Center, 1990.

Dossey, John A. "One Point of View: Elementary School Mathematics Specialists: Where Are They?" *Arithmetic Teacher* 32 (November 1984): 3, 50.

Johnston, Joseph, Jr., and Associates. *Those Who Can: Undergraduate Programs to Prepare Arts and Sciences Majors for Teaching.* Washington, D.C.: Association of American Colleges, 1989.

Mathematical Association of America, Committee on the Mathematical Education of Teachers. *A Call for Change: Recommendations for the Mathematical Preparation of Teachers of Mathematics.* Washington, D.C.: The Association, 1991.

——. *Guidelines for the Continuing Mathematical Education of Teachers.* Washington, D.C.: The Association, 1988.

Mathematical Association of America, Committee on Undergraduate Programs in Mathematics. *Recommendations on the Mathematical Preparation of Teachers.* Washington, D.C.: The Association, 1983.

National Council of Teachers of Mathematics. *Curriculum and Evaluation Standards for School Mathematics.* Reston, Va.: The Council, 1989.

——. *Mathematics Leaders in Elementary/Middle Schools.* Position statement. Reston, Va.: The Council, 1984.

——. *Professional Standards for Teaching Mathematics.* Reston, Va.: The Council, 1991.

National Research Council. *Everybody Counts: A Report to the Nation on the Future of Mathematics Education.* Washington, D.C.: National Academy Press, 1989.

U.S. Department of Education, Office of the Assistant Secretary for Educational Research and Improvement. *School Mathematics: Options for the Future.* Chairman's Report of the Conference. Washington, D.C.: U.S. Government Printing Office, 1984.

14

Learning How to
Teach via Problem Solving

Frank K. Lester, Jr.
Joanna O. Masingila
Sue Tinsley Mau
Diana V. Lambdin
Vânia Maria Pereira dos Santos
Anne M. Raymond

Curricula and instruction in our schools and colleges are years behind the times. They reflect neither the increased demand for higher-order thinking skills, nor the greatly expanded uses of the mathematical sciences, nor what we know about the best ways for students to learn mathematics.

—National Research Council, *Everybody Counts*

Instead of the expectation that skill in computation should precede word problems, experience with problems helps develop the ability to compute. Thus, present strategies for teaching may need to be reversed.... Students need to experience genuine problems regularly.

—National Council of Teachers of Mathematics,
Curriculum and Evaluation Standards for School Mathematics

Mathematics and mathematics education instructors in preservice and continuing education programs should model good mathematics teaching.

—National Council of Teachers of Mathematics,
Professional Standards for Teaching Mathematics

The course described in this paper was developed by a team at Indiana University—Bloomington led by Frank K. Lester, Jr. This paper was originally two papers, each describing different aspects of the course. One of the original papers was written by the first three authors, and the other by the latter three. Other members of the course-development team were Ronald Benbow, Carol Fry Bohlin, Philip Gloor, Michele LeBlanc, Francisco Egger Moellwald, and John Willems. The course development was supported by National Science Foundation grant no. NSF-TEI-8751478 to the Indiana University Mathematics Education Development Center, John F. LeBlanc, project director. The views expressed here are those of the authors and do not necessarily reflect those of the National Science Foundation.

IT IS time to change—and change drastically—the way mathematics is taught in our nation's schools, colleges, and universities. The three documents quoted above propose new views of learning and teaching that, if implemented, would result in profound changes in mathematics instruction. Herein lies a tremendous challenge for mathematics educators. To recommend that teachers adopt a drastically different posture toward mathematics instruction in their classrooms is one thing; to help them learn how to change is quite another. It is no small task to alter ways of thinking about and doing mathematics that have been bred by a dozen or more years of transmission-mode instruction. Simply put, we believe that teachers are prone to teach the way they themselves were taught. If elementary school teachers are led to think of mathematics as an arcane collection of rules, facts, and formulas that can be learned only by listening to lectures, completing routine worksheets, and endlessly practicing skill-building exercises, then we can hardly expect them to be enthusiastic about accepting an unfamiliar alternative.

Significant changes in the way mathematics is taught in our schools will come about only if significant changes occur in the way mathematics is taught in college. The National Research Council (1989, p. 65) agrees:

> Teachers themselves need experience in doing mathematics—in exploring, guessing, testing, estimating, arguing, and proving—in order to develop confidence that they can respond constructively to unexpected conjectures that emerge as students follow their own paths in approaching mathematical problems.

In this paper we describe a course developed over a period of three years at Indiana University to meet head on the challenge of the needed change. We begin with a brief discussion of the philosophy underlying this course.

TEACHING VIA PROBLEM SOLVING

In the 1989 Yearbook of the National Council of Teachers of Mathematics (NCTM), Schroeder and Lester (1989) discuss an approach to mathematics teaching that they call "teaching via problem solving." For them, when a teacher employs this approach, "problems are valued not only as a purpose for learning mathematics but also as a primary means of doing so. The teaching of a mathematical topic begins with a problem situation that embodies key aspects of the topic, and mathematical techniques are developed as reasonable responses to reasonable problems" (p. 33). Teaching via problem solving is an approach that embodies three key recommendations of the *Curriculum and Evaluation Standards for School Mathematics* (NCTM 1989): (1) mathematics concepts and skills should be learned in the context of solving problems, (2) the development of higher-level thinking processes should be fostered through problem-solving experiences, and (3) mathematics instruction should take place in an inquiry-oriented, problem-solving atmosphere.

Fundamental to teaching via problem solving is a change from viewing teaching as an act of transmitting information to passive students to an act of helping students construct a deep understanding of mathematical ideas and processes by engaging them in doing mathematics: creating, conjecturing, exploring, testing, and verifying. This change in point of view requires a correspondingly fundamental change in the teacher's role in the classroom. Rather than serve as the ultimate authority and dispenser of knowledge, the teacher variously plays the roles of guide, coach, question asker, and cosolver of problems (sometimes all these roles at the same time).

TEACHING TEACHERS TO TEACH VIA PROBLEM SOLVING

At Indiana University, we find ourselves in each of these roles as we teach our mathematics content course for elementary school preservice teachers (PSTs). In our experience, we can encourage the PSTs to become more mature mathematically by teaching via problem solving. In order to develop this maturity, the teachers of this course give special emphasis to four practices: (1) helping PSTs become reflective about their own mathematical behavior, (2) assisting PSTs in building mathematical connections, (3) engaging PSTs in cooperative learning, and (4) fostering an alternative view of assessment. In the following sections, we discuss each of these four emphases. Of these four, we discuss reflectiveness in more detail because we believe it to be a unique feature of our course.

Encouraging Reflectiveness

Perhaps the most important goal of our mathematics course is for the PSTs to develop adult-level insight into, and understanding of, the mathematics content appropriate for the elementary school. Central to the development of an appreciation of appropriate content is assisting PSTs in becoming more reflective about their own mathematical thinking and activity. As individuals engage in the process of making sense of mathematics and connecting mathematical ideas and topics, they gain mathematical understanding and, we hope, become aware of their own limitations in mathematics. Vital to the development of mathematical maturity is the cultivation of a disposition toward reflective thinking. Yinger and Clark (1981) have noted that through reflection students can learn (*a*) what they know, (*b*) what they feel, (*c*) what they do (and how), and (*d*) why they do it. To this list we would add two more items—that students can learn their strengths and weaknesses as doers of mathematics and that they can establish a measuring stick with which to measure their mathematical progress.

When PSTs are encouraged and expected to be reflective in their work, they become better at thinking reflectively, their understanding of the content improves, they are more creative and insightful in their problem solving, their motivation for learning increases, and they begin to look for and make connections among mathematical concepts. Students also begin to ask the types of questions routinely asked by mathematicians: What if this were ...? Does the generalization still hold? Why doesn't this case fit the pattern? Does this work in general?

PSTs can be encouraged to reflect as they work to understand a problem, evaluate their solution processes, decide if their solutions make sense, justify their generalizations, connect mathematical concepts, understand a problem solution different from their own, extend a problem, monitor their thinking processes, and communicate their ideas to other students and the teacher.

We have found several ways to encourage PSTs to be reflective: (*a*) having an experienced problem solver model how she monitors her thinking during the solution of a problem, (*b*) acting as a facilitator in the classroom rather than a dispenser of knowledge, and (*c*) having students engage in reflective writing and concept mapping. Each of these ways is discussed in the paragraphs that follow.

Modeling metacognitive awareness during problem solving

In the same way that children learn how to be adults by mimicking and interacting with adults, novice problem solvers become expert problem solvers by interacting with experts and with each other. Schoenfeld's (1985) distinction between expert and novice problem solver is characterized by the self-monitoring postures taken by each. Not surprisingly, Schoenfeld's experts are those who monitor most efficiently, whereas the novices are those whose monitoring activity is minimal at best.

Numerous researchers have tried to pinpoint just how students develop these monitoring activities. Hirabayashi and Shigematsu (1987) have suggested that the internal monitor developed by many students may actually be an internalized teacher who has influenced the student's mathematical activity. We believe that this idea of internalizing previously observed activity is important in the construction of a mature mathematical self. One way to give students an opportunity to develop an internal monitor is for them to watch an expert solve a problem. This expert, with the help of an external monitor, can demonstrate his or her metacognitively aware problem-solving posture.

All too often, problems are presented during the course of a lecture with nice, neat solutions carefully worked out ahead of time by the teacher. We have seen far too many students give up in the process of solving a problem because of the mistaken notion that all problems can be solved in a very short time. It is important for students to know that blind alleys and confusion are a normal part of problem-solving activity. By bringing experts into the classroom to talk aloud while they solve a problem, we help students become aware that even experts do not necessarily have all the answers at their fingertips. Experts, too, must stop to think when they encounter a real problem.

One of our PSTs, Pat, had an insight after seeing an expert model metacognitive activity while solving a problem in front of the class: "I realized that even mathematicians don't solve all of the problems they are faced with instantly. What a sneaky way to show us that it's okay to be frustrated and wander a little bit."

Vital to this modeling activity is the work of another individual, whom we call an "external monitor." The monitor has two roles: (1) to make certain that the problem solver does not ignore important information and (2) to force the problem solver to justify his or her solution strategy. During the problem-solving activity, the monitor models questioning techniques that help in developing internal monitoring, the hallmark of an expert problem solver.

An important component in encouraging the development of mature mathematical thinking is the class discussion that follows the modeling episode. It is important to encourage students to think about how the turning points in the expert's work may have come as a consequence of the monitoring that occurred during the problem-solving activity. In-class discussions can help students develop internal monitoring skills that will facilitate their successful problem solving and eliminate their (incorrectly perceived) need for the teacher's guidance. To help students develop a sense of their own problem-solving behaviors, it is also important to discuss the students' own experiences as problem solvers. This discussion affords an opportunity for students to become aware of their own thinking, planning, execution of strategies, and evaluation techniques (or the lack of them). Discussions throughout a course can also involve such topics as how long to go down blind alleys before rethinking a plan (Schoenfeld 1987) and how to evaluate the successfulness of various potential strategies.

Most important, students may need to be convinced to persevere. As any expert problem solver knows, real problems (those whose solution path is not immediately known to the problem solver) take time to solve. It is important for students to realize that successful problem solving requires time for understanding, planning, and evaluating, as well as time for carrying out the plan. Part of the process of developing mathematical maturity is learning to understand that good problem solvers continually monitor their own progress by asking themselves the questions that beginning problem solvers often rely on the teacher to ask.

Acting as a facilitator

For many elementary school children, the teacher is a very important person, often ranking second only to parents in importance. As a result, teachers play a key role in the development of their students' conceptualization of a mathematical self. Teachers who give detailed directions may be sending the message that students are to be dependent on the teacher for all knowledge. That is, teachers who tell the students everything they need to know restrict the development of student-initiated activity that is a natural consequence of students' curiosity, and they encourage students to take a passive role in their own education.

Students who exhibit behaviors that "increasingly assume control of the learning process" are developing autonomous learning behaviors (ALBs) (Meyer and

Koehler 1990). The development of ALBs is crucial if students are to become independent of the teacher as the authority figure. If the creation of ALBs is part of the development of a mature mathematical self—and we believe it is— teachers must adopt a balance, sensing when to intervene and redirect students' exploration and when to allow students to stumble along. This task is not easy. And yet, it is what we must ask of teachers, in particular, elementary school teachers.

At the college level, students no longer see the teacher as the second most important person in their lives. However, after years of schooling, they often have internalized that the teacher is the authority and that the teacher's word is as close to "the truth" as is possible. As a result, many of our college students, including those in teacher education programs, expect the university teachers to tell them what to learn and how to learn it. That is, they have not developed the ALBs (the autonomy from *their* teacher) that we want their future students to develop.

In the development of a mature mathematical identity, the teacher's role cannot be that of an authority. Rather, the teacher's role must be closer to that of a guide. The teacher chooses which problems to use as a means for introducing material and guides the discussion of these problems, but the teacher does not pronounce solutions. This is somewhat contrary to what has been considered appropriate teacher behavior in the past. Some learning theories advocate teacher behavior that models techniques and solutions so that, in mimicking these behaviors, the student will learn the material. Unfortunately, this "monkey-see, monkey-do" behavior rarely leads to a deeper understanding of mathematical principles and even more rarely leads to the development of autonomous learning behaviors.

The overarching responsibility of teachers is to establish a mathematical community in the classroom—a community where everyone's thinking is respected and where students' active engagement in mathematical activity is the norm. Within this community, the teacher's insightful questioning can play an important role in stimulating students' thinking so that they have opportunities to seek clarifications, strategies, and verifications without the teacher's direct intervention; to have their misconceptions challenged; and to examine and question their beliefs about mathematics.

Engaging PSTs in writing and mapping

Outside the classroom, PSTs can continue their reflective activity through reflective writing. Borasi and Rose (1989) have found evidence that reflective writing has four benefits for students: (1) therapeutic value, (2) increased learning of content, (3) improvements in learning and problem-solving skills, and (4) change in conceptions of mathematics. Reflective thinking and writing are "meaning-making processes that involve the learner in actively building connections between what she's learning and what is already known" (Mayher, Lester, and Pradl 1983, p. 78).

In general, we have developed two broad types of reflective-writing assignments for our course that give PSTs the opportunity to reflect on their feelings and thoughts and communicate them in written form. The following example is of the first type, which encourages the students to come to terms with their feelings and beliefs about mathematics:

- Discuss how you think mathematics fits with the real world.

Stacie responded by relating a personal experience:

> Personally I didn't think math fits very well with the real world—at least not my world. However, my view is beginning to change. I used to think only multiplication, addition, subtraction, and division had practical uses. Now I see myself using other ideas as well. For example, I wanted to paint stencils on my wall. I wanted them to be evenly spaced apart. I measured the area I wanted to paint, reduced it by four, figured out the spacing and enlarged it again. Before this class, I never would have thought to simplify the problem by reducing the area to be painted. I would just have eyed the spacing and hoped it turned out right. I used to think math was just numbers. Now I see that it is also concepts and numbers don't necessarily have to be involved.

The second type of reflective-writing assignment asks the students to discuss a mathematical idea or procedure. It forces the students to ask themselves, "Do I really understand this?" Two examples of open-ended assignments of this type are listed below:

- Make up and explain a divisibility test for 6 in base six.
- If p is prime, why is it true that p^7 has exactly eight factors? Would x^7 have exactly eight factors if x is a composite number?

Cathy responded to the assignment concerning the number of factors for prime and composite numbers in the following reflection:

> A prime number, p, has exactly two factors, 1 and p. If you square it, then it has 3 factors—1, p, and p^2. And, p^7 would have 8 factors—1, p^2, p^3, p^4, p^5, p^6, and p^7. This is because since p is prime, you are only adding one more factor when you increase the exponent by one—it is just p times the previous p^n. Therefore, p^n has $n + 1$ factors. For a composite number, x, x^7 would not have exactly 8 factors because the factors of x must be considered along with 1, x, x^2, x^3, x^4, x^5, x^6, and x^7.

Throughout a mathematics course, reflective-writing assignments furnish a "record of the writer's development through time, which can by itself provide new awareness and stimulus for reflection" (Borasi and Rose 1989, p. 353). This record also allows the teacher to enter into a dialogue with the student by responding to, challenging, and encouraging reflectiveness.

Another vehicle that promotes reflectiveness is the construction of concept maps. A concept map is a representation that elaborates a concept by connecting the concept with other mathematical topics, objects, and ideas. For example, using the word *fraction* as the original concept, a web of meaning could be developed that would illustrate one person's understanding of the concept of fraction. Other words or phrases would be linked with the original concept to demonstrate the relationship among the ideas. An example of a concept map, using the word *fraction* as the key concept, is shown in figure 14.1. The concept map and the accompanying reflection was constructed by Hannah, one of our PSTs, at the end of a unit on rational numbers.

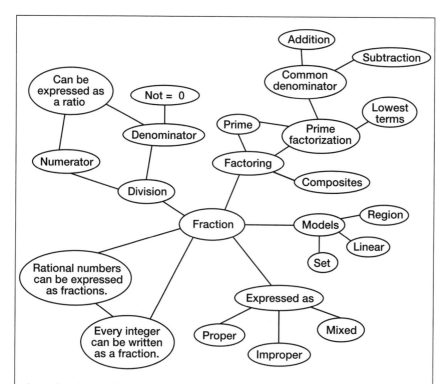

I am beginning to see more connections between fractions and other things.... I had just never seen them as all related. For example, previously I had never really thought much about the fact that in finding common denominators and putting fractions in lowest terms that prime factorization is an important concept. The different models we used to illustrate fractions— region models, linear models, and set models—also helped me look at fractions from a new perspective. I still don't have everything as organized in my mind as I'd like, but putting my ideas down on paper (in the concept map) helped me to organize some things.

Fig. 14.1. A preservice teacher's "fraction" concept map

Having students construct concept maps allows the teacher to see the students' thinking about a particular concept, but more important, it encourages the students to be reflective in organizing their thinking about the concept and in making connections among ideas related to this concept. (For more information about constructing concept maps, see Novak and Gowin [1984]).

Reflectiveness plays an essential role in the development of mathematical maturity. Without reflectiveness, students have no grounds for comparing their current mathematical maturity to their desired maturity. Just as "reflection sometimes produces a revelation" (Noddings 1984, p. 108) when considering one's relationship to others, through reflection students may begin to see their relationship to mathematical ideas and to gain a clearer picture of the mathematical maturity they would like to develop. We believe that this revelation is the foundational role of reflection in developing mature mathematical thinking.

Helping Students Build Mathematical Connections

The development of mathematical maturity through reflectiveness is aided by the building of connections among different ideas and topics in mathematics. Schroeder and Lester (1989) noted three ways that an individual's understanding in mathematics increases as connections are built among mathematical ideas: "(1) as he or she is able to relate a given mathematical idea to a greater number or variety of contexts, or (2) as he or she relates a given problem to a greater number of the mathematical ideas implicit in it, or (3) as he or she constructs relationships among the various mathematical ideas embedded in a problem" (p. 37).

In our course we choose problems that have significant, unifying mathematical ideas embedded within them, that is, ideas that pervade several areas of mathematics and serve to make connections among them. For example, the problem below, adapted from Thompson (1976, p. 79), is used to introduce the number-theory unit and to engage students in generalization and decomposition—two key mathematical ideas.

> Students at an elementary school tried an experiment. When recess was over, each student walked into the school one at a time. The first student opened all of the first 100 locker doors. The second student closed all the locker doors with even numbers. The third student changed all the locker doors with numbers that are multiples of 3. (*Change* means closing lockers that are open and opening lockers that are closed.) The fourth student changed the position of all locker doors numbered with multiples of 4; the fifth student changed the position of the lockers that are multiples of 5, and so on. After 100 students had entered the school, which locker doors were open?

Our students typically solve this problem by trying simpler cases (e.g., 30 lockers) and then forming a generalization. The process of decomposition (in this case, prime factorization) surfaces as the PSTs discuss and reflect on their solution process. Follow-up activities focus on constructing a general classification of the prime decomposition of integers with exactly one to ten factors. For example, integers with exactly three factors have a prime decomposition of p^2, where p is prime; integers with exactly four factors can be represented either by the decomposition p^3 or p^1p^2, where p, p^1 and p^2 are prime.

Promoting Cooperative Learning

Becoming a part of the mathematics community (being a learner and doer of mathematics) is an integral part of the development of mathematical maturity through reflectiveness. Working with others toward common goals, being actively involved in doing mathematics, and discussing and refining mathematical ideas are all ways of being a part of the mathematics community and can be accomplished through cooperative learning in small groups

Johnson and Johnson (1990) have said that cooperative learning *must* be used in mathematics classrooms. Higher achievement, better learning of mathematical concepts and skills, and gains in confidence in mathematical abilities are among the reasons offered to support this claim. Cooperative learning has been generally accepted for use in elementary school classrooms. We believe it is equally valid and essential at the college level. Besides the advantages noted by Johnson and Johnson, we believe that cooperative learning benefits students in the following ways: (*a*) students experience multiple approaches to solving a particular problem, (*b*) students see a variety of thinking strategies modeled by their small-group members, (*c*) groups can tackle more challenging problems than an individual could, (*d*) cooperative activities may yield more insightful work and stimulate further thinking, (*e*) working with others motivates students to persevere, (*f*) students develop their communication and reflection skills through the social interaction that a small group provides, and (*g*) the teacher's role changes from a dispenser of knowledge to a facilitator of cooperative learning efforts.

Students can be involved in cooperative learning situations in several ways. Perhaps the most obvious way is to have students work in small groups during class time (although cooperative work need not be limited to in-class work). Cooperative activities in which students work through negotiation to understand and solve problems and then evaluate their problem-solving efforts help move students toward autonomous learning behaviors. The students become less dependent on acquiring mathematical knowledge from an authority and more willing to take an active part in constructing their own knowledge

Lisa, one of our PSTs, noted the following when writing about working cooperatively:

> The group work has really made me think about what I'm thinking and writing. This fact is evident both when I'm trying to defend my answer to the skeptics in my group and when trying to explain my answer in writing. Group work also helped increase my perseverance.

Lisa's comments demonstrate the potential of cooperative work to enhance group members' reflectiveness and increase their perseverance.

Cooperative learning in the mathematics classroom can create an atmosphere where active learning is encouraged and expected not only by the teacher but also by the students. This atmosphere offers students a chance to become part of the mathematics community and to work toward developing mature mathematical thinking.

Developing a New Vision of Assessment

Because teaching via problem solving requires a different type of pedagogy and new and different expectations of students, it also demands new ways of assessing students' growth. In the past, assessment has often been thought of as synonymous with testing, and the results of assessment were frequently used primarily for grading. But teaching mathematics in the spirit of the *Curriculum and Evaluation Standards for School Mathematics* (NCTM 1989) necessitates a broader conception of assessment.

Several reasons can be cited for using more than just tests and quizzes for assessment in a mathematics class for prospective teachers. One, of course, derives from the belief that assessment should be embedded in classroom work and should be aligned with classroom methods. Thus, for example, it makes sense that group assessment would be used in a class where group problem solving is the norm. A second reason for using alternative assessment techniques is that the use of a variety of methods can provide a much richer impression of what students think, believe, and know than that obtained from any single method alone. Finally, in a course for teachers, it is especially important to model the use of alternative assessment. As a result of a dozen or more years of schooling, many students have developed the notion that the most important forms of assessment yield grades and serve to differentiate students from their peers. Ultimately, we would like to develop prospective teachers who have a broader view of assessment and who understand that in the long run, grades are less important than how an individual's understanding is being deepened. In the following paragraphs, we discuss several assessment techniques that we have used.

Classroom observation and interaction

Classroom observation and interaction can be used as an alternative form of assessment. If students solve problems and discuss mathematical ideas in small groups, an instructor can watch and listen carefully while circulating from group to group. A more complete picture of students' understandings can be constructed from a compilation of such observations than from any batch of test papers; as a result, more appropriate decisions concerning future instruction can be made.

Instructors can also gain important insights about the climate in their classroom as they interact with small groups of students: Are students confident? Frustrated? Involved? What beliefs about mathematics are being fostered by the work that students are doing? It is important to take time to think about how students feel and what they believe about the mathematics they are learning. Furthermore, when an instructor poses thought-provoking questions to a group of students working together and demands responses that give clear explanations of the students' reasoning, students soon learn the importance of considering why mathematics makes sense rather than focusing solely on answers. The students begin to appreciate the necessity of being able to communicate using precise mathematical language. Thus, the use of observation and interaction as assessment tools not only benefits the instructor but also helps students develop an appreciation for what is really important in the learning and teaching of mathematics—an appreciation that is particularly important if the students are prospective teachers who are to become reflective practitioners.

Although the use of a variety of assessment techniques helps instructors make decisions and aids in communicating their expectations to students, a major reason for assessing students' work is to judge progress and assign grades. Grades in our mathematics course for teachers are based in part on sources that seem to be quite traditional (e.g., quizzes, tests, and class participation) and in part on other, more obviously nontraditional aspects of the course, such as group presentations, written reflections, and concept maps. Yet even quizzes and tests may need to be assessed in new and different ways when the methods of teaching are nontraditional.

Group problem solving during tests

The *Curriculum and Evaluation Standards* stresses the importance of aligning assessment techniques with teaching methods. Since much of the work in our mathematics course for teachers is done in cooperative groups, tests consist of both individual problems and group work. The group portion of each test involves two phases. In Phase 1, students work in small, previously assigned groups to solve a problem and to write a single group solution. In Phase 2, individuals are expected to answer questions about the group's solution and to solve extensions of the problem. (See Kroll, Masingila, and Mau [1992] for a more detailed description of how cooperative work on tests can be graded.)

Having a group portion on each test emphasizes to students several underlying messages of the course: that mathematics is not a solitary endeavor, that there are a variety of alternative approaches to problems, and that clarity of communication is important. Moreover, group problem solving affords an opportunity for students of every ability level to work together and to contribute to a common goal. The assessment of problem solving in groups also makes the point that prospective teachers need to consider themselves responsible not only for their own learning but also for that of the others in the group. They should already consider themselves teachers.

Group presentations and group projects

Other sources of assessment data are group projects and presentations. In working on group projects outside of class, we have found that many students find themselves doing mathematics for the first time without continuous monitoring by a teacher. After having worked cooperatively to complete a project, the students communicate to their peers and their instructor (in a group report to the class and in individual written reflections) their self-assessments of the mathematics concepts they used and the difficulties they encountered in accomplishing the project. In their presentation to the class, students must, in a teacherly fashion, explain concepts and field questions about their mathematical thinking during the task. Thus, group presentations give instructors new layers to add to their multidimensional assessment of students' understanding and also furnish students still another situation in which they must engage in reflection and self-assessment.

Notebooks

Preservice teachers in our mathematics course keep a notebook that includes pages of daily activities, homework assignments, and quizzes, as well as concept maps and reflective-writing assignments.

The notebooks serve two assessment functions: (1) they help students become more aware of their own strengths and weaknesses, and (2) they provide an avenue for confidential one-to-one communication with the instructor. Two reflective writings by the same student written near the beginning and at the end of the semester show her growth as a learner and as a future teacher.

> *22 January.* I am so frustrated today. I am afraid I can't handle this class. We had to formulate the definition of a circle—I can't do things like this! I like the "traditional" math where they give me a problem and I have a formula that I can answer it with. I feel so stupid in this class.

> *26 April.* I'm getting ready to walk into our last class. This past semester there were times when I hated this class.... Something very special has happened to me this semester. All my life I have hated math with a passion. Now that I've had this class I can't believe that I once felt that way. This class has taught me that math has meaning behind it—it is not just doing step-by-step procedures with numbers. I am so glad that I took this class before I started teaching. Now math is exciting, and I will try my best to keep more students who don't like math from walking out of my classroom feeling that way.

CONCLUSION

A mathematics course for prospective teachers should help them develop more mathematical power (the ability to do mathematics and to have insight into the learning of it) but at the same time build a bridge to the development of teaching power (the ability to reflect on the complexities of learning and of designing and assessing instruction to facilitate it). Through a course with emphases such as the one we have described, students begin to reflect on the types of questions that good instructors ask to provoke students' thought and growth, the types of assignments that enhance students' thinking and learning, and—perhaps most important—the types of feedback that both students and teachers find beneficial. Prospective teachers' experiences with learning mathematics—which begin with their earliest years of schooling, are expanded and nurtured throughout teacher preparation, and are enhanced through self-assessment experiences—form the seeds for their implicit theories of mathematics and mathematics pedagogy. In turn, these implicit theories will grow and shape and inevitably be shaped by the teaching and assessment methods that these teachers of tomorrow choose to use in their day-to-day practice.

We began this paper by insisting that it is time to change the way mathematics is taught in our nation's colleges and universities. We have suggested that a good place to begin to make changes is in the mathematics content courses for preservice elementary school teachers. In line with this suggestion, we have described a course that engages PSTs in doing mathematics through solving problems, making and testing conjectures, and, in general, taking charge of their own learning. But as good as we think this course is, it is not enough! Much more must happen. We need a complete reconceptualization, a total overhaul, of the way elementary school PSTs are taught college mathematics. It is time for action. The longer we are willing to legitimatize college teaching that we believe is inadequate, the longer we will continue to find it difficult to prepare teachers who can give children the kinds of educational programs they truly deserve.

REFERENCES

Borasi, Raffaella, and Barbara J. Rose. "Journal Writing and Mathematics Instruction." *Educational Studies in Mathematics* 20 (November 1989): 347–65.

Johnson, David W., and Roger T. Johnson "Using Cooperative Learning in Math." In *Cooperative Learning in Mathematics,* edited by N. Davidson, pp. 103–25. Menlo Park, Calif.: Addison-Wesley Publishing Co., 1990.

Hirabayashi, Ichiei, and Keiichi Shigematsu. "Metacognition: The Role of the 'Inner Teacher.'" In *Proceedings of the Eleventh Annual Meeting of the International Group for the Psychology of Mathematics Education*, vol. 2, edited by J. C. Bergeron, N. Herscovics, and C. Kieran, pp. 243–49. Montreal: The Group, 1987.

Kroll, Diana Lambdin, Joanna O. Masingila, and Sue Tinsley Mau. "Cooperative Problem Solving: But What about Grading?" *Arithmetic Teacher* 39 (February 1992): 17–23.

Lester, Frank K., Jr., and Diana Lambdin Kroll. "Implementing the *Standards:* Evaluation: A New Vision." *Mathematics Teacher* 84 (April 1991): 276–84.

Mayher, John S., Nancy B. Lester, and Gordon M. Pradl. *Learning to Write/Writing to Learn.* Upper Montclair, N.J.: Boynton/Cook Publishers, 1983.

Meyer, Margaret R., and Mary Schatz Koehler. "Internal Influences on Gender Differences in Mathematics." In *Mathematics and Gender,* edited by Elizabeth Fennema and Gila C. Leder, pp. 60–95. New York: Teachers College Press, 1990.

National Council of Teachers of Mathematics. *Curriculum and Evaluation Standards for School Mathematics.* Reston, Va.: The Council, 1989.

―――. *Professional Standards for Teaching Mathematics.* Reston, Va.: The Council, 1991.

National Research Council, Mathematical Sciences Education Board. *Everybody Counts: A Report to the Nation on the Future of Mathematics Education.* Washington, D.C.: National Academy Press, 1989.

Noddings, Nel. *Caring: A Feminine Approach to Ethics and Moral Education.* Berkeley, Calif.: University of California Press, 1984.

Novak, Joseph D., and D. Bob Gowin. *Learning How to Learn.* New York: Cambridge University Press, 1984.

Schoenfeld, Alan H. *Mathematical Problem Solving.* San Diego: Academic Press, 1985.

―――. "What's All the Fuss about Metacognition?" In *Cognitive Science and Mathematics Education,* edited by Alan H. Schoenfeld, pp. 189–215. Hillsdale, N.J.: Lawrence Erlbaum Associates, 1987.

Schroeder, Thomas L., and Frank K. Lester, Jr. "Developing Understanding in Mathematics via Problem Solving." In *New Directions for Elementary School Mathematics,* 1989 Yearbook of the National Council of Teachers of Mathematics, edited by Paul R. Trafton, pp. 31–42. Reston, Va.: The Council, 1989.

Thompson, Maynard. *Number Theory.* Reading, Mass.: Addison-Wesley Publishing Co., 1976.

Yinger, Robert J., and Christopher M. Clark. *Reflective Journal Writing: Theory and Practice.* East Lansing, Mich.: Michigan State University, Institute for Research on Teaching, 1981. (ERIC Document Reproduction Service no. ED 208 411)

15

Preparing Secondary School Mathematics Teachers

A. Susan Gay

STUDENTS completing a bachelor's degree followed by a fifth year for certification bring to the teaching profession a solid background in content, preparation in pedagogy, and various field experiences. In support of the goal to make the education of teachers intellectually sound, the Holmes Group has recommended the development of graduate professional programs in teacher education (Holmes Group 1986). At the University of Kansas, the development of an extended teacher education program in the early 1980s was the result of a comprehensive self-examination. The increased responsibilities facing classroom teachers, additional requirements for teacher preparation from the state education agency, and a recognition that the undergraduate program was not adequately preparing teachers to function as professionals were noted as factors that prompted the program evaluation (University of Kansas 1990a). It was found that the four-year program alone did not furnish enough opportunity to educate preservice teachers in the knowledge base of the profession or supply enough clinical experience to develop professional practice (University of Kansas 1990a). Therefore, a five-year program was initiated.

The first students completed this extended teacher education program in 1986. The extended program includes more courses in mathematics and professional education as well as increased field experiences. Additional courses in arts and humanities as a part of the general education curriculum reinforced the belief in the importance of a strong liberal arts education for teachers (Holmes Group 1986; University of Kansas 1990a). This article will discuss the major components of the five-year teacher education program at the University of Kansas. An overview of the program is given in figure 15.1.

MATHEMATICAL BACKGROUND

Students majoring in the teaching of mathematics at the secondary school level complete approximately forty semester hours of mathematics, beginning

167

PROGRAM OVERVIEW

Freshman and Sophomore Years

- Courses in mathematics
- Courses in general education
- Course on introduction to teaching (includes field experience)
- Course on studying children and adolescents (includes field experience)

Junior and Senior Years

- Courses in mathematics
- Courses in professional education
- Field experiences each year

Bachelor of Science in Education Degree Completed

Certification Year

- Six-week student-teaching experience
- Graduate courses in professional education
- Development of a proposal for an action-research project
- Fourteen-week internship
- Completion of action-research project

Fig. 15.1

with calculus. The study of topics in analysis is completed with a capstone course in intermediate analysis, a rigorous study of concepts in calculus. Other mathematics coursework includes the study of linear algebra, modern algebra, modern geometry, statistics, and probability. Moreover, students select a course that focuses on mathematical modeling or other aspects of mathematical applications. Furthermore, courses in the area of computer science emphasize programming techniques and topics in discrete mathematics. Providing the teacher of mathematics with a view of mathematics as an organized body of knowledge would include the study of the historical development of mathematics.

The content areas recommended by both the Mathematical Association of America (Leitzel 1991) and the National Council of Teachers of Mathematics (1991) for the preparation of secondary school mathematics teachers, including number, geometry, analysis, discrete processes, probability and statistics, and mathematical structures, are present in the mathematical preparation program. Even though the foundations for a strong preparation in mathematics are present in the program, change continues to occur. An advisory committee on K–12 mathematics education for the department of mathematics includes not only university mathematicians but also mathematics educators at the university, junior college, and public school levels. This committee works to shape the college mathematics curriculum to meet the needs of future classroom teachers. The input from public school educators is essential in keeping the mathematical preparation of teachers current with changing emphases in the K–12 classrooms. Junior colleges are assuming an increasing role in providing the fundamental mathematical preparation of many university students. A working relationship between mathematics faculty members in junior colleges and those in the university is beginning. The initial stages are establishing lines of communication to work toward better coordination.

Cooperating teachers and principals consistently comment on the strong mathematical preparation the program gives students. These informal comments are consistent with evaluation results from a survey of principals who, as a group, maintained that new teachers from this five-year program were well prepared in the content areas (University of Kansas 1990b). The students themselves feel prepared mathematically as they begin their student teaching and their first year as new teachers.

GENERAL AND PROFESSIONAL EDUCATION

Students begin the study of professional education with a broad general education background of approximately sixty semester hours in language arts and communication, behavioral and social sciences, arts and humanities, science and mathematics, and personal and community health. In the junior year, students begin professional education coursework. For example, the study of human learning and development is the focus of a course on theories of human learning, motivation, and psychological and physical development that are relevant to the educational process. Understanding the educational needs of exceptional learners is one goal for another course that helps preservice teachers differentiate normal from atypical patterns of behavior and build a positive attitude toward atypical children and adolescents. A course in the social foundations of education takes a historical approach to the major social and philosophical foundations of American education.

Students from all subject areas work together in courses on general curriculum and classroom management. Teaching reading in the content area is a course for students majoring in teaching secondary-level subject areas. All students choosing middle or secondary school certification take a course in curriculum and instruction, which focuses on the specific issues and trends in middle-level education. Younger adolescents are distinctly different from older students. It is important for preservice teachers who will be working with grades 5–9 or 7–12 to have an understanding of the structure of middle schools and some experience with instructional strategies appropriate to young adolescents.

Students from all subject areas, at both the elementary and the secondary school levels, interact with one another in a course that develops an awareness of, and sensitivity to, multicultural education processes and that provides a rationale for a multicultural perspective within schools. A course on the use of instructional technology, including microcomputers and video technology, provides students with information on selecting instructional materials and media and opportunities to develop and produce a variety of instructional materials. Students completing the undergraduate program have acquired approximately thirty semester hours in professional education. In a survey of recent graduates of the program, more than 80 percent of them indicated that they found these required education classes valuable (University of Kansas 1990b).

Field Experiences

During the freshman year, most students with an interest in a career in K–12 education take a course that is an introduction to teaching. A field-based component of this course gives students eighteen hours of observing and aiding in a classroom, furnishing a realistic exposure to teaching. The purpose of this first field experience is to encourage the students to think about themselves as teachers and to increase their awareness of the roles of teachers.

Field experiences are a part of several other courses during the remainder of the undergraduate program so that students can spend time in public school classrooms each year. During the sophomore year, students spend time with children at different levels while enrolled in a course on studying children and adolescents in schools. The preservice teacher spends one hour each with a student in a junior high school classroom and a student in a senior high school classroom. Whereas these observations of secondary school students are arranged by the university, the preservice teacher must also spend one hour each observing a child at the preschool and elementary school levels, inside or outside a school setting. The purpose of these field experiences is to provide an introduction to the psychological study of children and their behavior.

During the junior year, university students usually take two courses that include field components. The introductory course on curriculum and instruction schedules two hours of observation and limited teaching each week for ten weeks at the preservice teacher's selected grade level. Students usually keep a journal during the course, writing about the different models of teaching and other observations they make in the classroom and relating aspects of the course to the classroom they visit each week. The course on education in a multicultural society allows students the opportunity to observe diversity within society by placing them in an urban or rural setting for eight hours of observation.

Two courses in the senior year continue to give the university students field experiences in their selected content area and grade level. One course focuses on classroom management, and four hours of observation is intended to give students opportunities to observe several models of classroom management. Two to four hours of observation of students in special settings accompanies the course that introduces the exceptional child. During the observation, the preservice teacher gains knowledge of, and direct experience with, exceptional students in an instructional setting.

Methods of Teaching Mathematics

The study of school mathematics and its interrelationship with specific instructional techniques is the focus of a three-semester-hour course in methods of teaching middle and secondary school mathematics. A wide variety of topics from general mathematics, algebra, and geometry is selected for study. The discussion of each mathematical topic includes various instructional techniques, concrete representations, applications, and connections, which can be used to provide an environment conducive to student learning. For example, components of the philosophy in NCTM's *Curriculum and Evaluation Standards for School Mathematics* (1989) are modeled as students experience mathematical problem-solving situations and discuss how they can be used to promote higher-order thinking skills, with different types of mathematical reasoning being used and with connections being made to other topics as well as to real-world experiences.

During the course, daily topics, such as decimals, systems of equations, and congruence, are scheduled for discussion. Working in pairs, students select a topic and design an instructional game or activity that could be used to introduce a concept or practice concepts in that area. Since all students in the course receive a copy of each activity, they begin to build a collection of the resources that they and their colleagues have developed.

New approaches to assessment are incorporated into the course to give the preservice teachers experience from the school students' point of view as well as an opportunity to analyze from a teacher's perspective. One example is the use of writing to assess student understanding. Writing, in paragraph form, how to solve a specific equation gives the preservice teacher experience with a writing task that may be used in a classroom. A class discussion analyzes the reasons why a teacher might use this type of task to detect student misconceptions.

Using real-world problems as group problem-solving projects is another form of assessment. Projects such as designing a chair or determining the seating arrangement for a restaurant pull together many of the concepts students have learned and give them a reason to use them. Students in the methods course participate in these projects and as a result develop experience with the project as an assessment tool.

The impact of technology on the mathematics curriculum is a frequent topic of discussion. Most of the students have seen middle and high school students who are dependent on calculators for the basic facts. They view negatively the use of calculators with fraction-operations, graphing, and symbolic-manipulation capability, seeing their use as additional steps toward student dependence on technology.

During the methods course, experiences with microcomputers are organized to expose students to a variety of types of software. Choosing from an instructor-generated list of mathematical software available in the educational computer laboratory, each student, using an evaluation form supplied in class, evaluates two different programs. One of the programs to be evaluated must be a graphing utility; the other may be a tutorial, game, drill-and-practice, or a problem-solving software program.

While studying geometric constructions, students spend one class period using a straightedge and compass as tools. The following class period is held in the computer laboratory, where students, using a geometry software package, work in pairs doing activities involving constructions. These experiences form a basis for discussing the use of computer software in mathematics and the importance of evaluating and selecting those programs that can be used to enhance learning.

The preservice teachers' reaction to using technology is a mixture of understanding the value of technology but appreciating the traditional approach to instruction with which they are most familiar. They have little experience with how technology can change the nature and emphasis of the mathematics curriculum. Discussions focus on pedagogical strategies, such as the kinds of questions teachers should ask, and on student activities, which can promote conceptual understanding.

Much of the course is designed to develop each student's self-image as a mathematics teacher. During discussions of issues in mathematics curriculum and instruction, students are encouraged to consider their own future classrooms and to see themselves as decision makers. On a regular basis, discourse centers on the sharing of ideas and experiences. This promotes reflective thinking by the preservice teacher, who begins to examine and consider the basis for making decisions and the consequences of those actions.

Through tutoring experience and field observations, most of the preservice mathematics teachers have worked with students who have a range of difficulties with mathematics. As various content topics, such as fractions, integers, solving equations, and geometric proofs, are discussed in the methods class, the students share the specific experiences they have had. Class discussion incorporates these examples and uses them to focus on theories of learning, such as constructivism, and theories of development, such as those of Piaget and the van Hieles. Sometimes methods students have worked with middle or high school students who have formed particular ideas about concepts or developed personal methods for working problems. The ideas of constructivism are directly related to these situations and can be discussed in the context of the examples described by the preservice teachers. Methods students may also be able to describe in terms of Piaget's theory or the van Hiele levels aspects of development seen in secondary school students they have tutored.

Students find the methods course to be particularly valuable. The resources from this class are used not only in the student-teaching experience but also during the first year of teaching. One student who had just completed his first year of teaching mathematics in a high school mentioned that as a new teacher he had few resources and indicated that the materials from his methods class were especially valuable to him. In a survey of principals who have employed graduates of the five-year program, administrators noted that the new teachers' teaching methods are good and that they are knowledgeable about teaching strategies (University of Kansas 1990b).

THE CERTIFICATION YEAR

The three major components of the certification year are a student-teaching experience, graduate coursework, and a teaching internship. For the first six weeks of the first semester, students complete a traditional student-teaching experience. The second component begins during the first semester and extends into the second semester. Building on foundations established in undergraduate experiences, the graduate courses focus on educational measurement in the classroom, the exceptional child in the regular classroom, the governance and organization of schools, and counseling and consultation skills for classroom teachers.

An important objective of the five-year program is to give students both an appreciation for educational research and experience using educational research methodologies to improve instruction and learning. As a part of their graduate coursework, students take a course in research-based analysis of teaching and learning. The course is designed to help students become knowledgeable about sources of professional literature and research findings and be able to use this information to guide their professional practice as educators.

Students from all disciplines take the same course, but activities and assignments are completed by the preservice teachers on topics of interest to them. During the course, the mathematics preservice teachers, for example, do a great deal of library work involving literature searches of research in mathematics education. They become familiar with journals and other publications that disseminate research and have experience critically reading articles from these publications. The preservice teachers learn how to analyze and synthesize research with special attention to how the research relates to classroom practice.

During the course, students gain experience in identifying and exploring an educational problem in mathematics education and designing a procedure to answer a research question. Ideas for action research have come from professional literature, the methods class, and the student-teaching experience. After selecting a topic of interest, the student develops an operational research question to guide the development of a research proposal. The written proposal begins with an introduction and rationale describing the research question, followed by a review of the literature related to the research topic. The last section of the proposal describes procedures for the research project, including sample selection, data collection, and analysis methods. At the completion of this course, each student has developed a proposal for an action-research project, which is conducted during the teaching internship.

In order to describe the process of conducting an action-research project, one student's work, which examined the use of manipulatives, is presented as an example. In her research proposal, the preservice teacher described research studies and recommendations that supported the use of manipulative materials in the middle school classroom. She proposed a quasi-experimental study that would involve four classes of middle school mathematics students. Two classes would be taught using manipulatives, and the other two classes would use paper-and-pencil methods. A comparison of measures on a pretest and posttest would determine whether differences existed between the two groups of students.

At the beginning of her internship after consulting with her cooperating teacher, the intern decided to use only two prealgebra classes and to use a two-week unit on the introduction of variables as the material to be covered during her research study. For this unit, the intern developed exercises that used algebra tiles to introduce and reinforce the uses of variables. A comparison of the pretest scores showed that there was not a significant difference between the classes prior to instruction. At the end of the unit, an examination of the means of the posttest scores indicated more improvement in the class that had used the manipulatives.

In discussing the results of her study, the intern speculated that the students might have been more comfortable using a familiar manipulative, rather than a new one, to learn the new concept. She did conclude, however, that the manipulatives did help the students develop an understanding of the concept of variable.

Other action-research projects have studied the effect of cooperative learning, writing activities, or the amount of homework time on student achievement. Teachers' attitudes about the use of different instructional strategies, such as peer tutoring and cooperative learning, have been investigated. The effect of cooperative learning, problem-solving activities, computer-assisted instruction, calculator use, feedback, and peer tutoring on students' attitudes have also been studied through action research.

The steps used to carry out the action-research project include submitting a proposal, making revisions, carrying out the research, and writing a paper describing the results. As a result of their experience with action research, new teachers develop not only research literacy but also a greater understanding of the importance of educational research in creating a learning environment. One aspect of professional development is exploring new ways to teach and experimenting with alternative approaches and strategies in the classroom (NCTM 1991). Those teachers with experience in action research have one more means of helping themselves identify effective teaching practices.

At the beginning of the second semester in the two weeks prior to the teaching internship, the students take a two-semester-hour course on instructional strategies in middle and secondary school mathematics. The purpose of this course is to promote the intern's self-understanding. Through reading assignments and discussions, the students analyze components of curriculum and instruction by examining worthwhile tasks, assessment techniques, classroom discourse (especially questioning strategies), and other aspects of maintaining a positive learning environment. The students draw from their field experiences, especially their student teaching, to formulate the broad framework as well as specific aspects of an individual philosophy of teaching. The fourteen weeks of the teaching internship complete the certification year.

Graduates of the program who have completed their first year of teaching mention the two teaching experiences as the most important components of the fifth year. For those students completing certification at both middle and secondary school levels, opportunities can be furnished at each level in one of the experiences. In comparing his preparation in a five-year program to that of others in four-year programs, one first-year teacher characterized himself as feeling "almost like having a year of classroom experience" before starting his first year of teaching.

Through research, Fuller (1969) described stages of concern teachers go through as they learn to teach. An early teaching stage is one filled with concerns of survival. It is possible that the students completing the five-year program spend less time at this stage, which is characterized by concern for personal adequacy in the classroom. With the student-teaching and internship experiences as foundations, many of them may move more quickly to later stages, where concern is focused on the teaching situation and the pupils.

Administrators have noted the confidence in graduates from the five-year certification program (University of Kansas 1990b). Principals credited the extended program and particularly the two teaching experiences with giving these new teachers more poise, confidence, and maturity than graduates of four-year programs at other institutions (University of Kansas 1990b). Principals also noted that these graduates have skills "beyond [those of] the typical beginning teacher" (p. 116) and "teach like experienced teachers" (p. 109).

More than half the students interviewed in 1989 who graduated from the five-year program believed that the program had a positive effect on their obtaining employment (University of Kansas 1990b). In the survey of alumni, significant differences between four-year-program alumni and five-year-program alumni were found in how graduates believed the program helped them meet job or career requirements. Ninety-five percent of the five-year group compared with 75 percent of the four-year group found the School of Education to be very effective or effective with respect to this goal of the program.

CONCLUSION

Since the beginning of the five-year teacher education program, the number of students completing preparation as secondary school mathematics teachers has increased from one or two a year to fifteen to twenty a year. These students bring a commitment to the profession because they see teaching as an important career choice, not as a second or part-time job.

The teacher education program continues to be assessed to identify areas where improvement can be made. One goal of the program is to help teachers improve instruction and student learning. The action-research project completed during the internship gives preservice teachers experience using research methodologies to improve instruction and learning. However, effective teachers are also those who are thoughtful about their practice (Porter and Brophy 1988). They take time to evaluate the success of different teaching strategies, materials, and curriculum plans in guiding student learning and behavior.

The course on analysis of teaching and learning is being broadened to include additional ways to analyze the effectiveness of teaching. It continues to be important for preservice teachers to be research literate, apply research findings to classroom practice, and use action-research methods to answer questions about teaching effectiveness. Reflection and self-evaluation can also be used to analyze the teaching and learning occurring in a classroom. The use of journals, case studies, and portfolios may be an effective way to pull together the preservice teacher's experiences, beginning with professional education coursework (including the methods course), continuing through student teaching and graduate coursework, and culminating in the internship experience.

Reshaping the program continues in an effort to meet the needs of preservice teachers. There are still questions to be answered about changes being proposed and implemented. The purpose of the program, however, remains unchanged: to provide first-year teachers with the foundation needed to become effective mathematics educators.

REFERENCES

Fuller, Frances F. "Concerns of Teachers: A Developmental Conceptualization." *American Educational Research Journal* 6 (1969): 207–26.

Holmes Group. *Tomorrow's Teachers: A Report of the Holmes Group.* East Lansing, Mich.: The Group, 1986.

Leitzel, James R. C. ed. *A Call for Change: Recommendations for the Mathematical Preparation of Teachers of Mathematics.* Washington, D.C.: Mathematical Association of America, 1991.

National Council of Teachers of Mathematics. *Curriculum and Evaluation Standards for School Mathematics.* Reston, Va.: The Council, 1989.

———. *Professional Standards for Teaching Mathematics.* Reston, Va.: The Council, 1991.

Porter, Andrew C., and Jere Brophy. "Synthesis of Research on Good Teaching: Insights from the Work of the Institute for Research on Teaching." *Educational Leadership* 45 (1988): 74–84.

University of Kansas, School of Education. *National Council for Accreditation of Teacher Education and Kansas State Department of Education Institutional Report, Volume II: Standards for the Unit.* Lawrence, Kans.: University of Kansas, 1990a.

———. *School of Education Follow-Up Studies, 1985–1990: A Longitudinal Appraisal of Program Outcomes Composite Report.* Lawrence, Kans.: University of Kansas, 1990b.

16

Problem-Solving Interviews as Preparations for Teaching Mathematics

Harriett C. Bebout

ROBLEM-SOLVING interviews create a potentially rich setting for preservice teachers to learn about mathematics learning and teaching. By observing a student's actions and listening to explanations during mathematics problem solving, a preservice teacher has the opportunity to look closely at the student's mathematical thinking strategies and processes. This knowledge about how students think mathematically provides an essential background for planning classroom instruction that will promote mathematical thinking and problem solving. Along with this primary goal of learning about students' mathematical thinking, the interviews furnish opportunities to focus on other important mathematics education issues, specifically those issues about equity and about teachers' beliefs about who can do mathematics and about what type of mathematics should be included in the school curriculum.

This idea of interviewing students is not new. Nearly forty years ago in a classic *Arithmetic Teacher* article, "Big Dividends from Little Interviews," Weaver (1955) described the benefits of interviewing students to help teachers develop an understanding of students' mathematical processes. Weaver's classic idea from the 1950s is especially relevant today as the mathematics education community addresses national standards for mathematics teaching, specifically Standard 3: "Knowing Students as Learners of Mathematics" (National Council of Teachers of Mathematics [NCTM] 1991, p. 144).

In addition to the use of problem-solving interviews in mathematics teacher preparation, individual interviews with mathematics learners have been used extensively by researchers to learn about the development of mathematical thinking. In each mathematical content area, most current researchers use some type of individual interview to identify, document, and study the development of mathematical knowledge in their particular content area. This same information is valuable especially for the mathematics teaching community, whose responsibilities are to plan instruction that will bring mathematics understandings alive in classroom situations. A place to begin to understand mathematical thinking is in problem-solving interviews during preservice professional development.

PROBLEM-SOLVING INTERVIEWS IN AN URBAN TEACHER EDUCATION PROGRAM

In an urban-based mathematics methods course, one of the early field assignments for preservice mathematics teachers is to carry out a mathematics problem-solving interview with a student who has been designated a low achiever by the teacher. Urban classroom settings are important to this specific population of preservice teachers who, for the most part, attended suburban or parochial schools prior to entering the university. Their field experiences in urban classrooms offer opportunities for reflection on their conventional ideas about mathematics learners and about school mathematics tasks. For example, many of these teachers-to-be have media-generated beliefs about urban students and their low levels of mathematics achievement; the assignment of the problem-solving interview is designed to help preservice teachers discover the mathematical power that students actually do possess when they are involved in meaningful mathematical tasks.

For the interview assignment, the preservice teachers are given the following brief guidelines:

1. Select a student who is not achieving well in mathematics.

2. Compose several problems (at least three) that are at a high level for the student; make up problems that relate to the student's daily out-of-school activities.

3. Supply appropriate materials or manipulatives that match each of the problems so that the student's concrete or pictorial representations will be observable; avoid problems that allow the student to represent only symbolically or that involve only symbol manipulation.

4. In the written report of the interview, (a) state the problem that was given to the student, (b) describe the student's solution processes or strategies for each problem, and (c) reflect on and identify the steps of Polya that were used by the student.

5. If possible, use a tape recorder so that you as interviewer can focus your attention on the student rather than on taking notes; the audiotape will help you reconstruct and reflect on the interview.

The interview assignment is designed to highlight specific goals of the elementary mathematics methods course. These goals include helping the preservice teachers (1) to learn about students' mathematical thinking, that is, to become aware of students' powerful informal or natural strategies that often are very inventive and very different from those of adults; (2) to identify the steps of problem solving, that is, to see the steps of problem solving in action, especially the first two steps of understanding and representing (Polya 1957); and (3) to reflect on issues of equity and conventional attitudes about mathematics, that is, to appreciate the previously uncovered mathematical powers of students who have been termed classroom low achievers and the potential negative results of labeling these students.

Learning about Students' Mathematical Thinking

Listening to, and observing the actions of, a student who is solving a mathematics problem in a one-on-one interview setting enables a preservice teacher to focus on mathematical thinking. Such a setting allows the preservice teacher to concentrate on the individual student by alleviating the concerns of classroom management. The setting also provides opportunities for the preservice teacher to observe a student's extended processes or solution paths rather than only the final product or answer. Seeing the solution processes helps the preservice teachers understand both that a student's thinking often is different from their own adult thinking and that real mathematics problem solving is more than a quick-recall process.

In the following excerpt from an interview assignment with a third-grade boy, Preservice Teacher A (PTA) discovered that the student used a different approach from the one she was expecting. She asked Student N the following problem:

PTA: You have 45 compact disks. You need to put them away in cases that hold 4 each. How many cases will you need?

Her notes of the interview with N follow:

After N read the problem, he put the blocks in groups of four and then counted the blocks by fours. He counted up to 40 and thought he had answered the problem. He read the problem again and realized that the question said he had 45, not 40, disks. He then made one more group of four and said, "11 cases." I asked him how many disks he had laid out and he said, "44." So I asked him to look at the problem again. He said, "It says 45 but there is only 1 disk. Can I put it in a case by itself or just leave it in the stereo?" I said to put it in a case. He then said that he would need 12 cases in all.

N used blocks as a concrete model to help him solve this problem. He represented the problem by grouping the blocks together in fours and then solved his problem by counting the groups. I thought he solved his problem in a backward manner. The problem said that a case would hold 4 disks and he had 45 disks to start with. He counted by fours, and then counted to 40. He caught his mistake when he went back and checked his answer by counting the individual blocks to make sure he had the right amount. He saw he had only 10 groups equaling 40 when he should have had 12 groups equaling 45. If I had done this problem, I would have counted out 45 and then grouped in fours. I wouldn't have thought of his way, but it worked for him. Both ways are right; neither way is wrong. It was so neat the way he focused on real life by wanting to leave the extra one in the stereo.

In the next interview, Preservice Teacher B (PTB) especially noted the mathematical power of a first-grade girl. The following problem was read to Student J:

PTB: If Julie drinks 2 glasses of milk every day, how many glasses of milk will she drink in one full week?

This problem was really tough for J. She first asked me, "Do I use plus or take away?" I told her to use whatever she thought would work. After a lot of thinking, she finally said, "14." I was really surprised and asked her how she got that answer. She said that she thought about a number line and counted by twos, and that for every two spaces she held up a finger until she reached seven fingers. When I asked her why she stopped there, she said, "Because there are seven days in a week." I was really impressed. The first graders never have any high-level or real thinking problems like this one in their books, and they never use any numbers higher than 10. I thought it was fantastic that she could actually solve a math problem like this, since she is not a very good student in learning her facts.

To say the least, I was intrigued by this assignment. Learning about children's thinking and how they solve math problems is really interesting. It never dawned on me that students could use so many different and unusual methods to solve mathematical problems. It displays a lot of intelligence for these kids to figure out problems when they've had no prior experience. I think that an interview can be very helpful too in locating problems in a child's thought processes. Once a problem is discovered, it is much easier to help a student having trouble with math. This was a great assignment because I learned a lot about children's thinking.

Identifying the Steps of Mathematical Problem Solving

In addition to developing a respect for student thinking, the individual interviews are important means for activating major ideas about mathematics problem solving. Specifically, these ideas include learning to differentiate between high- and low-level mathematical tasks and then learning to identify the steps of problem solving when a student is engaged in a high-level task.

Learning to tell the difference between low- and high-level mathematical tasks, that is, between problems that call for the recall of facts or algorithmic procedures and those that call for more creative thinking, is an important concept for future mathematics teachers. By observing a student's solution processes during an interview, a preservice teacher can begin to sense the differences in thinking that are generated by low- and high-level problem types. For the most part, if a student has a ready answer or a memorized algorithm or solution path, then the problem is a low-level problem for that student; however, if the student has no ready answer or solution path and has to search for a way to solve it, then the problem is a high-level one for that student. Such insights into the selection of high-level problem types can be fostered by observing student thinking during the interview.

Learning to identify the steps of mathematical problem solving, that is, Polya's (1957) structure of understanding, representing, solving, and checking, is another important concept for preservice mathematics teachers. By observing and listening to a student who is solving a high-level problem, the preservice teacher often can see Polya's structure evolve: First, the student goes through a process of understanding or clarifying the problem; next, the student searches for a way to represent, or show, the problem by using concrete, pictorial, or symbolic models or a combination of models; then, the student uses the representation to solve the problem; and finally, if the correctness of the solution is important enough, the student reflects on the reasonableness of the answer. The interview setting presents trial-and-error experiences with real student problem solvers so that preservice teachers can begin to appreciate the major role that problem solving plays in the mathematics curriculum and in mathematics instruction.

In the following two problems, Preservice Teacher C (PTC) identified the problem-solving steps of a second-grade boy (S):

PTC: A squirrel has 45 nuts. He wants to hide them evenly in 7 different places to be sure another squirrel does not get them. How many nuts will the squirrel put in each hiding place and will there be any left over?

S: Wait! Tell me the story again.

PTC: (Repeats the problem.)

S: Can I pretend to hide them for myself?

PTC: (Nods.)

S: (Counts the nuts out into 7 piles and has 3 left; counts the number in each pile.) There are six in the piles, and I get to eat three of them now.

PTC: OK. Good. Now, how about this one: We are going to have a party and we are going to make chocolate chip cookies. If there are 96 chocolate chips in the whole bag and we need to make 24 cookies, how many chips should we put in each cookie to keep them fair or even? Will we have any chocolate chips left over?

S: What are we going to do again?

PTC: (Repeats the problem.)

S: (Counts out 24 cups; then puts a poker chip in each cup while counting to 96; looks in all the cups.) Four chips are in each cookie and nobody will get cheated.

What I learned from doing this interview was that when we think the students are not catching on, that is a bad assumption. It is probably that the problem doesn't make sense to them. For example, the boy I interviewed asked me to repeat each problem at least once. Even after I repeated the problem, it took him a few minutes to begin work. I also learned that it is natural for a kid to go through those four steps in solving a problem. He had to understand the story before he could begin to set it up, and then, once it was set up, he checked to be sure the question was what he thought. When he was finished, he went over it to be sure it was right. I didn't tell him to do these steps; they just came naturally. I think that's amazing.

In the next interview, Preservice Teacher D (PTD) was interested especially in the representations used by a fourth-grade girl:

PTD: Tasha wants to order an ice-cream sundae. But it is tough to make a decision about what to order, since she can choose between three kinds of ice cream and five toppings. She can have vanilla, chocolate, or strawberry ice cream.

Tasha can top her sundae with caramel, hot fudge, marshmallow, strawberry, or butterscotch. How many possible sundaes are there if Tasha can
choose one ice cream and one topping?

*L read this problem through a couple of times and just sat at the table, not knowing
where to begin. Finally she began talking out loud. "OK," she said, "vanilla could have
caramel, hot fudge, marshmallow, strawberry, or butterscotch." She chose a white circle
(I had different-colored big and small geometric pieces on the table) to be her vanilla
and put five different-colored shapes around it. She then did the same for the chocolate
and strawberry ice creams. She told me there were fifteen possible combinations. When I
asked how she came to this answer, she said that she had "added all the fives because
each kind of ice cream could have five different things on top."*

*I was really surprised and pleased with L's success in solving this problem. L used the
problem-solving steps to her advantage here. I think she understood the problem and
what it was asking right away but had more difficulty representing the problem. I think
she wanted to use manipulatives but wasn't quite sure how to go about it. Solving the
problem once she was able to represent the ice cream and toppings came rather easily. In
fact, I think she was quite confident in her answer. However, she did check her answer to
some extent by reading over the problem again to see if her representation made sense.*

*Through this interview and the others, too, I would have to say that the key to problem
solving is in representing the problem. L solved the problem (after understanding it) by
using concrete materials. She must have asked herself, "What would this problem look
like?" Through the representation she was able to come up with the correct answer. An
interesting note is that the only problem she missed was one that she represented the
wrong way. As a result of these interviews, I would have to generalize that students in a
classroom should be guided and maybe even taught the importance of representing the
problem—either concretely or by drawing a picture. Representation seems to be the key to
success, and helping students to feel successful is the name of the math game, so to speak.*

Reflecting on Issues of Mathematics Equity and Conventional Attitudes

Interviewing students in problem-solving situations is also a way to address
with preservice teachers some of the conventional attitudes about mathematics
learners and the mathematics curriculum that make equity in mathematics education a continuing concern. These attitudes are discussed in both the general
literature on educationally disadvantaged or marginal students (Sinclair and
Ghory 1987) and the specific literature on mathematics and underrepresented
student groups (Secada 1992). Briefly, many of these conventional attitudes
involve specifying the groups of students who can and cannot succeed in mathematics and describing the type of mathematics that should be taught to the less
successful students. Such conventional attitudes are present in many preservice
teachers as well as in their veteran counterparts.

By carrying out interviews in urban schools with students from groups who are
underrepresented in mathematics and mathematics-related careers (typically,
schoolchildren from low-income families), preservice teachers have opportunities
to see the mathematical thinking strategies of these students and consequently to
reflect on conventional attitudes about mathematics in relation to students from

these groups. Very often, the performance of a student in an interview setting is very different from that student's classroom performance and from the expectations stated, implicitly or explicitly, by the cooperating teacher. Listening to individual students and observing their strategies is a way to help preservice teachers thoughtfully examine their attitudes and beliefs.

In addition to learning about the mathematical thinking of students who are labeled unsuccessful in mathematics, problem-solving interviews are a means for examining the type of mathematics that is recommended for these students. Some conventional approaches suggest that students must first learn basic skills before being able to do high-level problem solving (Doyle 1991; Knapp and Turnbull 1991). In contrast, the interviews give preservice teachers an opportunity to see these students performing well in high-level problem-solving situations. Essentially, preservice teachers often learn from their interviews that even though a student may not know basic facts or algorithms, either through a lack of interest or a lack of memorization, that student can become motivated to solve interesting and relevant high-level mathematics problems composed for the interview. Often, the preservice teachers begin to understand that a student is mathematically literate whether or not that student has memorized basic facts and algorithms.

In the following interview excerpt with a third-grade girl, Preservice Teacher E (PTE) discussed her assumptions about low-achieving students:

PTE: It took Mr. Cobb 1/4 hour to walk to the supermarket. He stayed there for 1 1/4 hours. It took him 1/2 hour to walk home carrying all his groceries. How long was Mr. Cobb gone?

At first the student did not know what 1/4 of an hour equaled. I explained that 1/4 hour and a quarter of an hour are the same. She asked if a quarter of an hour was 15 minutes. I answered affirmatively, and then she proceeded to work out the problem. First she wrote down in familiar terms of minutes what each fraction of time was equal to. Then she added the two 15-minute times and circled them to remind herself that she had used them already. Next, she added the two 30-minute blocks of time and came up with 60 minutes. She immediately knew that 60 minutes equals one hour and added the two one-hour blocks of time. Her final answer was 2 hours.

I asked the student to show me if her answer was correct by using a clock with movable hands. She said, "I need to know what time he started." Before I could answer, she said, "I'll just pretend that he started at 12:00." First she moved the hands of the clock to 12:00. Then she moved the minute hand to 12:15. Next she moved the minute hand completely around to 1:15, stopped and made sure the hour hand had moved to the 1 and proceeded to move the minute hand forward another 15 minutes. Then she moved the minute hand the final 30 minutes and announced that he finished at 2:00. She said, "12 to 1 is one hour and 1 to 2 is two hours. I was right." And of course she was.

This exercise made me realize just how easy it is to assume that a low achiever can't do the work. However, by taking the time to watch her work out a problem, I could see her thinking processes through her actions. I realized that she can do the work but for some reason she isn't. This makes me question why she isn't staying interested in math. Is she bored? Does she need more manipulatives? Does she need more teacher patience? Does she need more practice? All the above? This interview showed me that we shouldn't give up on the low achievers, but we should try just that much harder to include and encourage them.

In the following excerpt from an interview with a fourth-grade girl (M), Pre-service Teacher F (PTF) reflected on teacher expectations:

> PTF: All the stores on Main Street are either white or green or brown. There are twice as many white stores as brown stores. There are 5 more brown stores than green stores. There are 7 green stores. How many stores are there on Main Street?

I had to help M read the problem. I ended up reading it to her and then asking her to retell it to me in her own words to be sure she understood. Then I gave M some paper stores to use, thirty green, thirty brown, and thirty white stores. Here's what we said:

> M: It all depends on the green stores.
> PTF: What do you mean?
> M: Well, because first I have to use the green stores to find out the brown, and then I need the brown to find out the white ones.
> PTF: OK, try it.

M counted out seven green stores and then wrote on her paper, "7 + 5 = 12"; she then counted out twelve brown stores. Then she wrote down "12 × 2 = 24" and counted out white stores by twos to reach twenty-four. Then she counted up the stores she had set out: 7 green, 12 brown, and 24 white. She counted by fives to reach forty and then counted by ones to reach forty-three.

According to her teacher, M is getting a D in math. Her teacher says she is really lazy and doesn't care about her work. But in my opinion, once I helped her with reading the problem, she was a very capable math student. M very easily went through the entire problem-solving process from understanding the problem to representing it to solving the problem (but I didn't see her check her answer). This interview along with the others I did has taught me to be cautious about accepting the ideas of other teachers that could be wrong about students. From my experience in the interview, I think M is actually a good student in math.

CONCLUSION

In summary, these interviews are very worthwhile for most preservice teachers. They learn specifically that most students have powerful informal or natural strategies that often are very inventive and very different from those of adults. They also come to see Polya's steps in action through identifying and understanding the problem-solving processes. They begin to appreciate too the previously uncovered mathematical powers of students who have been termed classroom low achievers and the potential negative results of labeling these students. In addition, the insights gained from the interview settings may be important in helping preservice teachers to examine more critically the conventional instruments used for mathematics assessment. As they begin to discover the wealth of strategies that students use and to appreciate the importance of processes over products in problem solving, these future teachers may no longer be content with product-oriented assessment instruments. Indeed, they may become major supporters of revised forms of evaluation that focus on mathematical processes.

REFERENCES

Doyle, Walter. "Classroom Tasks: The Core of Learning from Teaching." In *Better Schooling for the Children of Poverty: Alternatives to Conventional Wisdom,* edited by Michael S. Knapp and Patrick M. Shields, pp. 235–55. Berkeley, Calif.: McCutchan Publishing Corp., 1991.

Knapp, Michael S., and Brenda J. Turnbull. "Alternatives to Conventional Wisdom." In *Better Schooling for the Children of Poverty: Alternatives to Conventional Wisdom,* edited by Michael S. Knapp and Patrick M. Shields, pp. 329–53. Berkeley, Calif.: McCutchan Publishing Corp., 1991.

National Council of Teachers of Mathematics. *Professional Standards for Teaching Mathematics.* Reston, Va.: The Council, 1991.

Polya, Gyorgy. *How to Solve It.* Princeton, N.J.: Princeton University Press, 1957.

Secada, Walter G. "Race, Ethnicity, Social Class, Language, and Achievement in Mathematics." In *Handbook of Research on Mathematics Teaching and Learning,* edited by Douglas A. Grouws, pp. 623–60. New York: Macmillan Publishing Co., 1992.

Sinclair, Robert L., and Ward J. Ghory. *Reaching Marginal Students: A Primary Concern for School Renewal.* Berkeley, Calif.: McCutchan Publishing Corp., 1987.

Weaver, J. Fred. "Big Dividends from Little Interviews." *Arithmetic Teacher* 2 (April 1955): 40–47.

17

Modeling the NCTM *Standards:* Ideas for Initial Teacher Preparation Programs

Michaele F. Chappell
Denisse R. Thompson

A S THE mathematics education community continues its efforts to understand the mechanisms of change for practitioners seeking to teach according to the recommendations of the National Council of Teachers of mathematics (NCTM) (1989, 1991), we also must consider the initial experiences of prospective teachers of mathematics. Such a consideration undoubtedly involves examining the requirements for preparing prospective mathematics teachers at the college and university levels. A major concern relates to the varied opportunities that undergraduate students are given in their initial preparation programs (*a*) to be aware of their growth as mathematics educators, (*b*) to perceive their students as mathematics learners, and (*c*) to understand the mathematics pedagogy appropriate for their students (NCTM 1991).

The learning environment that teacher education students experience is, perhaps, the most important factor of the students' entire educational program. Such environments can greatly affect students' beliefs and attitudes about how mathematics should be taught in the various grade levels. Even the prospective teacher who understands and is sensitive to the need for changes in the mathematics curriculum and in pedagogical techniques occasionally experiences moments of skepticism, frustration, and pure disbelief. Thus, a strong rationale exists for university faculty to create, as often as possible, supportive classroom situations that allow the prospective teacher to perceive the complexities involved in teaching mathematics to individuals.

Most educators believe that teaching is a complex process involving a blend of content knowledge, pedagogical knowledge, and the ability to make appropriate instructional and professional decisions. Certainly, mathematics education courses need to assess prospective teachers' content knowledge. However, because teaching involves more than having a strong content base, we have felt a need to include additional experiences that allow our students to understand other aspects of a teacher's role. These experiences give the prospective teachers a foundation on which they can build during their initial years as practitioners in the classroom.

This paper presents as exemplars appropriate assignments that we have used in courses for prospective elementary, middle, and secondary school teachers in order to offer them an array of diverse professional development opportunities. We do not imply that the methods or assignments are unparalleled elsewhere in initial preparation programs. On the contrary, it is probable that other teacher education programs have established the kinds of assignments (or variations of them) described in this article as assessments in fundamental courses for preservice students in all disciplines.

CREATING VIABLE ASSESSMENTS

We view the assignments described in this paper as models of alternative assessments in the courses, so that prospective teachers can begin to think about alternative means of assessing students' learning in mathematics classrooms. The assignments are outlined and discussed in three broad categories: (1) assignments that focus on the instructional role of the teacher; (2) assignments that focus on the teacher as a decision maker; and (3) assignments or activities that help to establish professional involvement. In each of these categories, we provide sample assessments.

The Instructional Role of the Teacher

Initial preparation programs must offer opportunities for prospective teachers of mathematics to learn how to use effective questioning techniques in their classrooms. Such opportunities allow prospective teachers to formulate appropriate questions that will investigate students' mathematical reasoning. In the mathematics classroom, such questions are often more difficult to generate than the typical questions requiring responses of yes or no or questions that elicit only one correct response. Becoming proficient at formulating probing questions that elicit divergent responses requires experience over time. Thus, engaging prospective teachers in assignments that will help them perfect their formal and informal questioning skills is advantageous.

As a prospective mathematics teacher begins the process of developing an identity as a teacher of mathematics, it is important that she or he learn how to reflect on and examine her or his instruction from various perspectives. Becoming a reflective practitioner involves a degree of self-awareness, particularly an awareness of how one interacts with students in various instructional contexts. Two assignments that allow undergraduates to reflect on their questioning techniques and interactions with participants are detailed in figure 17.1, Clinical Pupil Study, and figure 17.2, Ministudy.

CLINICAL PUPIL STUDY

Description: This assignment requires you to conduct a Clinical Pupil Study with a child or small group of children (maximum of 3). Listed below are some guidelines to follow:

1. This assignment is to be conducted with a partner. It is extremely beneficial to conduct such an assignment in collaboration with your colleagues. Designing an interview-oriented study and analyzing its results with a partner afford an opportunity for both of you to understand the data from perspectives other than your own.

2. Select a particular mathematics activity that interests you. Either choose the activity from a textbook, article, or class discussion, or create an original exercise. Plan the activity to last fifteen to forty-five minutes, depending on the age of the children.

3. Your activity must be approved. To obtain pre-approval, submit a summary briefly describing the activity, the children (age or grade level), any materials to be used, the objectives of the activity, and general plans for conducting the interview. Include some of the questions you expect to ask during the interview. Remember that in order to understand the students' thinking, these questions will have to require more than yes or no responses.

4. In the interview, you and your partner are to observe carefully and ask thought-provoking questions to help you understand the students' mathematical activity. Because you are interested in how the students develop mathematics concepts, do not encourage right responses and discourage wrong ones. Your goal is to design interview questions that will get at the students' thinking strategies.

5. All interviews should be audiotaped so that you and your partner will have a record of the children's responses and of the activity. Videotaping is permissible. Be certain that the taping equipment is in proper order (e.g., batteries are charged, correct buttons are pressed) before interviewing.

6. As soon as possible after completing the interview, you and your partner should debrief each other and prepare a report on the students you interviewed. In the report, include ages, grades, and any pertinent information you obtained from the students. The report should be typed, and double-spaced in paragraph form, and it should be a minimum of three pages in length. The sentences should be well formed and grammatically correct.

7. In the interview report, you should describe the children's mathematical activity or responses. Attempt to infer their reasoning. It will be useful to include in your report specific dialogue segments to support your statements. Some specific questions to answer in the report include (but are not limited to) the following:

   a) What observations were made of the children's activity?
   b) What explanations were given by the children for their activity?
   c) What diagrams and figures, if any, were made by children?
   d) What explanations do you make of their responses?
   e) What conclusions, if any, can you make about the children's learning of mathematics?

Evaluation: Grading criteria:

a) Initial summary (see items 2 and 3)	10
b) Conducting interview (see items 1, 4, and 5)	10
c) Completeness of the report (see items 6 and 7)	25
d) Grammar and format (see item 6)	5
Total Points	**50** (out of a course total of 400 points)

Fig. 17.1. Clinical Pupil Study used in elementary school methods courses

MINISTUDY

Description: One special assignment for this class is to conduct a mathematics learning Ministudy of a youngster (or a group of youngsters). Depending on your topic and its structure, this assignment will involve various degrees of field-based observations, individual or group interviews, and a curriculum unit written for a particular mathematics topic. A comprehensive analysis of the students' mathematics learning is to be submitted as a result of the study. Some specific guidelines for the assignment are as follows:

1. For your Ministudy, select a mathematics topic taught at the secondary school (7–12) level that interests you. Your Ministudy should last between three and five days. That time gives you a chance to collect substantial data for later analysis. Although the days do not have to be consecutive, they probably should be less than a week apart. The study should take the form of sequential lessons, each ranging from forty-five to sixty minutes in length.

2. Try to choose a topic that can be enhanced with problem-solving activities and manipulative materials. Focus on a concept that could be taught within a three-to-five-day period. Sample topics might be finding areas with tangrams, rational-number addition with pattern blocks, or integer operations using two-color chips.

3. As part of each lesson, you should ask probing questions to help you understand the students' mathematical reasoning. You will have to design different questioning techniques that will get at students' thinking strategies. Remember, you are interested in how the students learn the concepts on the basis of their previous experiences.

4. Your Ministudy should be approved. Submit a summary of your study briefly outlining the topic, the grade level of the individuals with whom you'll be working, the sequential lessons, any materials to be used, the objectives of the lessons, and general plans for conducting the study. Also include some of the questions you expect to ask during the study. To get at the students' thought processes, these questions will have to require more than yes or no responses.

5. In order to analyze your data properly, it is essential that you at least audiotape all your lessons. If possible, videotape them (particularly if students are to be working with hands-on materials). Be certain that the taping equipment is in good order (e.g., batteries are charged, recorder is on) before beginning each lesson. In some instances, you may have to seek the parents' or guardian's permission before taping students.

6. A final report of your Ministudy is required. It should consist of a comprehensive analysis of your data. In the final report, include ages, grades, and any pertinent information you obtain from the students. Also include a description of the students' mathematical responses. Attempt to infer their reasoning. It is useful to quote students' verbal reactions to support your statements. Some specific research questions to ask within any study include (but are not limited to) the following:

    a) What observations were made of the students' activity?
    b) How do students learn the topic with problem-solving activities or manipulative materials?
    c) What explanations were given by the students for their activity?
    d) What explanations do you make of students' responses?
    e) What conclusions, if any, can you make about the students' mathematics learning?

7. The final report should be typed and double-spaced in paragraph form, and it should be a minimum of five pages in length. The sentences should be well formed and grammatically correct.

8. You will have a chance to share with the class the results of your study through a ten-to-fifteen-minute presentation.

Evaluation: Grading criteria:

a) Initial summary (see items 1 and 4)	10
b) Conducting study (see items 2, 3, and 5)	25
c) Comprehensive analysis (see item 6)	40
d) Grammar and format of the report (see item 7)	10
e) Presentation (see item 8)	15
Total Points	**100** (out of a course total of 400 points)

Fig. 17.2. Ministudy used in secondary school methods courses

The Clinical Pupil Study has been assigned in a series of two mathematics education courses taught to prospective elementary school teachers. The Ministudy assignment has been used in a mathematics education course geared to teaching mathematics in the middle grades. Although different assignments are required of the two groups of prospective teachers, the assignments produce some common benefits.

First, the prospective teachers can learn how to assess students' knowledge informally through carefully designed sequences of tasks or through carefully sequenced questions posed in an interview. Included in the tasks created for the Clinical Pupil Study are questions that primarily seek to understand the students' mathematical responses rather than simply determine if they can "get the right answer." Second, the prospective teachers can obtain more information through the interviews than through a paper-and-pencil test about what, how, and why the student knows; thus, they can learn and appreciate the value of *listening* to students. In the Clinical Pupil Study, prospective teachers focus more on sequencing tasks within a particular lesson (i.e., the interview) and are therefore primarily concerned with short-range planning issues.

Similarly, for each Ministudy lesson, prospective teachers have to phrase several questions that seek to understand how students are connecting ideas across the various lessons. Additionally, the Ministudy requires prospective teachers to think critically about their role as teachers of mathematics. The teacher education students must prepare a sequence of several lessons that are designed to develop or enhance students' learning of a particular mathematics concept. Further, they collect data about the effectiveness of the lessons. This is an initial step toward developing reflective practitioners capable of conducting research in their own classrooms. This assignment directly involves prospective teachers in long-range planning, since they are concerned primarily with tasks that are presented across the lessons.

In addition to assignments that focus on questioning strategies, teacher education students need to do assignments that focus on making instructional decisions. Figure 17.3, Activity File, and figure 17.4, Reading and Writing Communication Project, detail two assignments requiring prospective mathematics teachers to develop routine and nonroutine activities that will help students master various mathematics concepts.

ACTIVITY FILE

Description: You are to create an Activity File that you can use when you begin your professional teaching career. The contents of your file should be appropriate for the grade levels you plan to teach. Other guidelines for the file are listed below.

1. The file should contain some *nonroutine* problems and activities, that is, problems and activities that are problematic, interesting, and mathematically rich. The problems and activities should be similar to the nonroutine problems and activities that we have discussed during the semester. However, the scope of the problems and activities certainly is much wider than what has been shared in class.

2. Your activities may be designed for use with an entire class, with a small group (a minimum of four students), or in a learning center. Overall, the file needs to target a range of grades (K–6). A specific activity can target a range of grades, or different activities can target different grades.

3. Your file should contain at least ten problem-solving activities that will promote higher-order thinking. You should have at least one activity in each of the following areas:

   a) Prenumber or place-value or grouping concepts
   b) Whole number addition or subtraction
   c) Whole number multiplication or division
   d) Number sense or number theory
   e) Rational number addition or subtraction
   f) Rational number multiplication or division

4. All materials for each activity should be available so that the activity is ready for use.

5. Prepare a single-spaced, typed one-and-one-half-to-two-page summary of all the activities in your file. For each activity, state the objectives, target group of children, materials, plans, evaluation, and so on.

6. You can use a variety of resources to find activities that are appropriate for use in a problem-centered, small-group, or whole-class environment.

7. Plan to present your favorite activity from your file.

Evaluation: Grading criteria:

a) Summary of activities (see item 5)	10
b) Completeness of file (see items 3 and 4)	25
c) Richness of activities (see item 1)	15
d) Creativity and variety of activities (see items 2 and 6)	15
e) Class presentation (see item 7)	5
Total Points	**70** (out of a course total of 420 points)

Fig. 17.3. Activity File used in elementary school methods courses

READING AND WRITING COMMUNICATION PROJECT

Description: Choose a topic from some area in the secondary school mathematics curriculum. Make the topic as focused as possible; one way to do so is to consider your topic as a typical unit of study, that is, as a chapter. Below are the guidelines you should follow:

1. Identify the vocabulary and mathematics symbols that are important in understanding the chosen concept.

2. Develop three activities that involve understanding vocabulary or notation.

    a) Two of these activities should be appropriate for the beginning of a unit to help students initially learn the vocabulary or notation.

    b) The third activity should be appropriate for a summary activity.

3. Write ten word problems or problem-solving scenarios related to this topic. Your problems should be realistic and interesting to students. They ought to be problems that students will be motivated to solve. Above all, avoid the contrived word problems that are typical of textbook problems.

4. Develop three writing activities for use in this unit.

    a) Two of these activities should be for use during the development of the unit.

    b) The other activity should be designed as part or all of an assessment.

Evaluation: Grading criteria:

a) Vocabulary activities (see item 2)	20
b) Word problems (see item 3)	20
c) Writing activities (see item 4)	15
d) Creativity	10
e) Completeness	5
Total Points	**70** (out of a course total of 350 points)

Fig. 17.4. Reading and Writing Communication Project used in secondary school methods courses

The Activity File has been assigned in a mathematics education course taught to prospective elementary school teachers. The Reading and Writing Communication Project has been used in a secondary school mathematics education course geared to reading in the content area. The two assignments share some benefits. In both assignments, the prospective teachers have opportunities to develop instructional activities that are appealing to students. The activities in the Activity File have to be developed for a broad range of content topics. However, in the Communication Project, the activities must be developed for use as initial or concluding tasks in a given content topic. Both assignments require that prospective teachers consider tasks that enhance instruction in the classroom.

Creativity is often a component of these assignments. Prospective teachers become aware of additional mathematics educational resources and learn to adapt ideas to meet their needs.

The assignments discussed in this section require several weeks to complete. Thus, they help prospective teachers understand the length of time that could be involved in instructional planning.

The Teacher as a Decision Maker

Practicing teachers make many decisions daily about curriculum-related matters. In the mathematics classroom, such decisions are frequently made while constructing tests. The Test Construction (see fig. 17.5) is an assignment requiring students to make decisions about a test and its appropriateness for two different groups of students.

TEST CONSTRUCTION

Description: This assignment requires you to design a test that would be appropriate for students at two different ability levels. Guidelines for this assignment follow.

1. Choose a chapter from a first-year algebra or geometry textbook. Design a chapter test that would be appropriate for use with a homogeneous class. Your test should adequately cover the content of the chapter and should reflect the skills and concepts that you expect students to master. To help you design the test, you should develop a set of objectives for the chapter.

2. Modify the test created in item 1 so that it would be appropriate for use with students in an honors section of the course. Although some items might be identical on both tests, attempt to include problems that reflect the mastery level expected of honors students.

3. Create an answer key for both tests.

4. Write a typed one-page rationale indicating the differences between the tests and why you think those differences are significant in assessing the knowledge of the students being tested.

5. You should submit the following:

   a) Bibliographic citation for the textbook on which the test is based
   b) Set of objectives for the chapter
   c) Both tests with answer keys
   d) Rationale

Evaluation: Grading criteria:

a) Set of objectives	10
b) Extent to which test covers objectives	10
c) Variety of test items	10
d) Appropriateness of items for homogeneous group	15
e) Appropriateness of items for honors group	15
f) Correctness of answer keys	5
g) Rationale	10
Total Points	**75** (out of a course total of 450 points)

Fig. 17.5. Test Construction used in secondary school methods courses

In addition to test construction, it seems only appropriate in initial preparation programs for prospective teachers of mathematics to evaluate and make decisions about curriculum-support materials and resources available for the mathematics classroom. Inherent in the Textbook Evaluation (see fig. 17.6), the Curriculum Strand (see fig. 17.7), and the Budget Analysis (see fig. 17.8) is the process of decision making in different contexts.

In most instances, the present mathematics curriculum is very different from the curriculum that prospective teachers recall from their own school experiences. It is extremely valuable for preservice teachers to know the extent to which resources (e.g., textbooks, supplementary materials) conform to what they are learning in the methods courses and to the NCTM's Standards. The Textbook Evaluation and Curriculum Strand assignments, in particular, force prospective teachers to rethink the breadth of the mathematics curriculum, the content, and the "scope and sequence" at the various grade levels. Teachers often have to give their opinion on spending funds for instructional purposes. The Budget Analysis enables preservice students to learn about the many varied manipulatives that are available for use in developing and modeling abstract mathematics concepts. Additionally, it requires prospective teachers to make difficult decisions about which materials are most useful at a particular grade level. By making choices and writing rationales supporting their proposed purchases, prospective teachers can understand better the decisions to select certain items over others and the factors (e.g., available budget, range of materials, intended use, single sets vs. classroom sets) that influence those selections.

TEXTBOOK EVALUATION

Description: The purpose of this assignment is to help you critically analyze a mathematics textbook to determine the extent to which it meets the content objectives and satisfies the recommendations of recent reports dealing with mathematics education. Some guidelines for the assignment are as follows:

1. Choose two textbooks from the secondary school curriculum that have a recent copyright (1990–present). One textbook should be a basic sixth-, seventh-, or eighth-grade mathematics textbook, and the other textbook should cover algebra or geometry.

2. Carefully analyze the content of the books, paying special attention to the extent to which one of the following topics is included:

   a) Technology, both calculator and computer

   b) Realistic applications

   c) Problem solving, including nonroutine problems and appropriate strategies

   d) Review of content covered in previous years

   e) Depth to which content is covered

   f) Variety and complexity of problems

3. For each textbook, write a typed three-to-four-page narrative summarizing your critique of the textbook. Your narrative should contain some indication of the extent to which you judge that the textbooks agree with such documents as the NCTM's *Curriculum and Evaluation Standards*. You should support your comments with examples from the textbook and rationales from the documents.

Evaluation: In grading the Textbook Evaluation, careful attention is given to the extent to which students address their chosen topic in item 2 and the extent to which they justify their comments. In particular, we look for evidence that the students consider whether the topic is truly integrated into the materials or included only as enrichment. That is, we want to see evidence that students have examined the materials critically and carefully. The total points for this assignment range from 20 to 25 points out of a course total of 400 points.

Fig. 17.6. Textbook Evaluation used in secondary school methods courses

CURRICULUM STRAND

Description: The purpose of this Curriculum Strand assignment is to help you become familiar with some of the new elementary school mathematics textbooks, particularly with the extent to which new content appears in the books. Two options exist for completing the assignment: Option 1 is to work in groups of three to four; option 2 is to work individually. Some guidelines for the assignment are as follows:

1. Your task is to take a curriculum topic from the list below and trace the development of that topic through the elementary school curriculum. (If you work in groups, you should trace the topic through grades K–6; if you work individually, you should trace the topic through two consecutive grade levels.) **Choices of topics require approval to ensure variety in topics and textbook series.**

Topics:

   a) Place value and whole number operations

   b) Rational number operations

   c) Measurement (metric and customary), including length, area, volume, capacity, and weight

   d) Plane (2-D) geometry, including perimeter and area of figures and transformations of figures

   e) Solid (3-D) geometry, including surface area and volume of figures

   f) Probability, statistics, graphs, and coordinate graphing

2. Select a current mathematics textbook series (copyrighted from 1990 to the present).

3. You should evaluate the extent to which the topic appears throughout the textbooks for the grades you are studying. You will need to determine *broad content objectives* (not behavioral objectives) for your topic and determine how those objectives are met in the textbook lessons. In particular, you must document the following:

   a) Whether or not the objective is covered at the grade levels you have selected to study

   b) The level of complexity of the objective at each grade (such as which figures are studied, which measurement units are studied, what types of graphs are studied) and how the objective is extended through the grade levels

   c) What kinds of activities, questions, and instruction are given to help students attain the objective

4. You need to submit two items:

   a) a narrative summary that includes—

      (1) a bibliographic reference indicating the textbook series you selected and its publisher, authors, and copyright date

      (2) a summary of your analysis of the extent to which your content topic is addressed in the series, including at what levels and with what approaches

      (3) an evaluation of the extent to which the series concurs with the NCTM's *Curriculum and Evaluation Standards*

   b) A documentation of your objectives. Each objective should be on a separate page. For each grade, address items *a, b,* and *c* in item 3 above. If an objective is not present at a particular grade, indicate "not present."

Evaluation: Grading criteria:

Option 1: Narrative summary (see item 4a)	10
Thoroughness (see items 3 and 4b)	25
Format (see item 3)	10
Group evaluation	5
Total Points	**50** (out of a course total of 400 points)
Option 2: Narrative summary (see item 4a)	15
Thoroughness (see items 3 and 4b)	25
Format (see item 3)	10
Total Points	**50** (out of a course total of 400 points)

Fig. 17.7. Curriculum Strand used in elementary school methods courses

BUDGET ANALYSIS

Description: You are given $500 at the beginning of the school year to spend on materials for your classroom. Create a Budget Analysis according to the guidelines below:

1. Determine a single grade level or a two-grade range. Those of you in special education may want to consider a slightly larger grade range.

2. Decide what manipulative materials you would like to purchase for your classroom. You may decide to buy classroom sets of items or only demonstration sets. Calculators are acceptable purchases, but you can spend no more than one-third of your money on them.

3. Prepare a request for a purchase order for the items. Include the item description, cost, catalog number, and supplier. Remember that schools are tax-exempt and do not pay sales tax. Also, do not worry about shipping charges; the school will absorb those costs. Try to get your totals as close to $500 as possible without going over the amount. Remember that no one likes to give back money that has been allocated.

4. Prepare a two-to-three-page rationale for your purchases. Incorporate some philosophical reasons for the selections you made, and state why the selected materials are essential for effective mathematics teaching.

5. This is your opportunity to become aware of existing manipulative materials and to think about what materials you would like to have in your classroom. The money should be spent for concrete materials or supplements. Assume that the school purchases the required textbooks.

Evaluation: In the evaluation of the Budget Analysis assignments, we carefully observe the list of items purchased by the preservice students and the amount of money spent on those items. The rationales for the purchases receive particular attention. We look for evidence that the students made prudent decisions about the materials purchased and for sound reasons that these materials are important in the teaching and learning of mathematics. The total points for this assignment range from 20 to 30 points out of a course total of 420 points.

Fig.17.8. Budget Analysis used in elementary school methods courses

Establishing Professional Involvement

The initial teacher preparation phase is, perhaps, one of the most opportune times for prospective teachers of mathematics to establish professional involvement. Getting involved in the activities and offerings of the mathematics education community can be achieved through several means. First, prospective teachers can be required to observe and participate in a classroom prior to the typical internship. Such experiences give the prospective teachers additional opportunities to interact with students as well as to observe, reflect on, and discuss classroom dynamics. The Field-Based Participant Observations assignment (see fig. 17.9) offers such an opportunity.

FIELD-BASED PARTICIPANT OBSERVATIONS

Description: The primary purpose of the Field-Based Participant Observations is to provide more internship opportunities to secondary school mathematics education students. As a result, you will have an opportunity to experience classroom dynamics related to content and pedagogy. You should note the following guidelines:

1. This assignment requires you to conduct fifteen to twenty fifty-minute observations during the semester. You should spread your visits over a seven-to-eight-week period. Try to observe two different grade levels or two different subjects within a grade level. Try to observe in the same classes so that you have some continuity and, perhaps, become familiar with the makup of the class.

2. As you observe in the classrooms, interact with the students and participate in their mathematics activities to understand how the students make sense of the mathematics.

3. Keep written records of your observations.

   a) Keep a journal of your experiences. Your journal should describe the lessons you observe and any activities in which you participate. Be specific. Why did you think the lesson went well or went poorly? What kinds of strategies did the teacher use during the lesson? How did the teacher handle discipline and behavior problems?

   b) Keep an observation log recording the time, date, and level of the class in which you observe. The teacher should initial your observation log after each visit.

4. During the observation period, you should complete the following activities:

   a) Work with an individual student, perhaps in a tutorial role or during homework time.

   b) Work with a small group of students (three to five).

   c) Grade a set of homework or classwork papers.

   d) Review homework or classwork with the students.

   e) Do an enrichment activity with the class using a game or some manipulative.

   f) Design a quiz, review it with the teacher, administer it, and grade it.

   g) Teach one lesson; you might choose to connect the quiz in (*f*) to the lesson you teach as a means of determining how much the students learn from the lessson.

   h) Interview the teacher about the school, his or her classes, the challenges and joys of teaching, and any other matters of interest to you; summarize your interview in a two-to-three-page paper.

5. At the end of your observation period, submit the following items:
  a) The journal and the observation log of your visits
  b) A portfolio describing each of the experiences in item 4. Be sure to include your quiz, your lesson plan for the lesson you taught, and your summary of your interview with the teacher.
  c) A summary (one to two typed pages) reflecting on the experience

Evaluation: Grading criteria:
  a) Completion of activities (see items 3b and 4)   30
  b) Thoroughness of journal (see item 3a)   30
  c) Quality of portfolio (see item 5b)   30
  d) Summary (see item 5c)   10
Total Points   **100** (out of a course total of 450 points)

Fig.17.9. Field-Based Participant Observations used in secondary school methods courses

Another means for building professional involvement is through Journal Abstracts (see fig. 17.10). This assignment requires the prospective teacher to select methods-oriented or research-oriented articles pertaining to the teaching or learning of mathematics. The students then write an abstract of the article including individual reflections on the strengths and weaknesses of the article. The Journal Abstracts have enabled prospective teachers to familiarize themselves with (1) the wide variety of mathematics education journals and (2) the research conducted on different topics within the field. Preservice students have gained insight into the ways of teaching particular mathematics topics, the ways people construct mathematics, and the problems that school students encounter in learning mathematics. Many of the preservice students begin to wonder and comment about why they were not taught mathematics by using more concrete approaches. Hence, the Journal Abstracts assignment confirms the validity of the content and methods used in the university classes and encourages the preservice students to use hands-on activities in their future classrooms.

Certainly, reading journals and writing about the related research are valuable. However, prospective teachers can also be enticed to become professionally involved through attendance at a state, regional, or national mathematics education conference.

CONCLUSION

In creating viable assessments for preservice teachers of mathematics, we aim to make certain that the assessments will engage our students in different kinds of activities. Although the time required to grade such assignments can be substantial, we believe that the benefits to the students are well worth our effort. We learn more about the benefits of these assignments as we continue to adapt them to meet the needs of our students. We encourage others to share our experiences.

JOURNAL ABSTRACTS

Description: The Journal Abstracts are designed to give you an opportunity to read some of the professional literature relating to the content of the course. Your abstract should follow the guidelines set forth below:

1. Abstracts should span two to four double-spaced, typewritten pages.

2. Include a complete journal citation—title, author(s), journal, volume, month or season and year, and inclusive pages.

3. Summarize the article in your own words. The summary should provide sufficient detail to give the reader a good understanding of the article.

4. In one to two paragraphs, give your reaction, critique, and opinion of the article, indicating where you stand on the methods or issues discussed in the article. That is, do you agree with the author and why or why not? How does the article relate to the issues and methods discussed in class?

5. You are responsible for three journal abstracts. Two of the articles you use should pertain to methods or issues in teaching mathematics. The remaining article should describe some research in mathematics education detailing a specific study, the participants used in the study, the design of the study, and the results.

6. Possible sources for articles are the *Arithmetic Teacher*, the *Mathematics Teacher*, the *Journal for Research in Mathematics Education*, and *Science and Mathematics*.

Evaluation: Each abstract is worth 10 to 15 points. We consider how well the article is summarized as well as the overall quality of the writing.

Fig. 17.10. Journal Abstracts used in elementary and secondary school methods courses

REFERENCES

National Council of Teachers of Mathematics. *Curriculum and Evaluation Standards for School Mathematics*. Reston, Va.: The Council, 1989.

———. *Professional Standards for Teaching Mathematics*. Reston, Va.: The Council, 1991.

18

Teachers and Students Learning Together in After-School Mathematics Enrichment Laboratories

Nadine S. Bezuk
Frank A. Holmes
Larry K. Sowder

THIS article discusses a model of teacher preparation that is based on the premise that *all* students, regardless of race, culture, gender, primary language, or economic situation, can learn mathematics. This view is echoed in the National Council of Teachers of Mathematics (NCTM) *Professional Standards for Teaching Mathematics* (NCTM 1991), which states that "every student can —and should—learn to reason and solve problems, to make connections across a rich web of topics and experiences, and to communicate mathematical ideas" (p. 21).

Educators need to consider ways of enhancing the mathematics education of all children, including the impact of role models and quality teaching on these children. As noted in *Everybody Counts* (National Research Council 1989, p. 21),

> During the next decade, 30 percent of public school children, but only 5 percent of their mathematics teachers, will be minorities. The inescapable fact is that two demographic forces—increasing Black and Hispanic youth in the classrooms, decreasing Black and Hispanic graduates in mathematics—will virtually eliminate classroom role models for those students who most need motivation, incentive, and high-quality teaching of mathematics.

Even though school populations are increasingly nonwhite (and expected to become 80 percent nonwhite in southern California), projections predict relatively few nonwhite preservice teachers. One estimate (Quality Education for Minorities Project 1990) is that only 5 percent of the teacher work force over the next decade will be nonwhite, a *decline* from the current 10 percent! Hence

The project described in this article was supported in part by the National Science Foundation. The opinions expressed are those of the authors and not necessarily those of the Foundation.

it is imperative that those African American and Hispanic preservice teachers "in the pipeline" be encouraged to continue to prepare to teach and that their preparation be as strong as possible

This article describes one project's efforts in encouraging ethnically diverse preservice elementary school teachers to continue pursuing teaching careers while strengthening their understanding of mathematics and their awareness of the components of effective mathematics teaching. These efforts are one purpose of the San Diego Mathematics Enhancement Project (SDMEP). The overarching goal of the SDMEP is to enhance the abilities of teachers in grades 2–8 to help all students, particularly those who are African American and Hispanic, learn mathematics. This is accomplished by strengthening and expanding teachers' knowledge of mathematics content and teaching methodology through their participation in a two-year-long cycle of in-service sessions, after-school mathematics laboratory activities with students, and workshops for parents.

The keystone of the SDMEP is the after-school mathematics enrichment laboratories, which are conducted by the project's in-service teachers with the assistance of the preservice teachers and are aimed at promoting students' mathematics achievement and disposition. These enrichment laboratories meet after school for three hours each week.

Preservice teachers—African American and Hispanic whenever possible—attend workshops with in-service teachers, for which they can receive course credit. Each preservice teacher is assigned to a single after-school laboratory group, where he or she not only observes and works with an experienced teacher of African American or Hispanic children but also plays an active role in the laboratory sessions, as, for example, in working with a small group on some instructional activity. Since this work takes place before the teacher education program (a fifth year in California, where only a minimal exposure to schools is required beforehand), it is particularly motivating to the preservice teachers and gives them an unusual opportunity to learn about mathematics teaching and learning in today's curriculum. These college students also serve as role models to the children in the laboratory sessions: "Here is someone like me, and he/she is going to college!"

Project workshops are based on the recommendations of NCTM's *Curriculum and Evaluation Standards for School Mathematics* (1989), the *Professional Standards for Teaching Mathematics* (NCTM 1991), and the *Mathematics Framework for California Public Schools* (California State Department of Education 1992). Although individual workshop sessions focus on the major content strands, pedagogical practices that may be new to the teachers in the project are woven through each of the content topics. In this way, these practices are revisited, strengthened, and expanded for the teachers over the course of two years. The major content strands that are covered include number, patterns and functions, algebra, statistics, probability, geometry, and measurement. Pedagogical aspects woven through these content themes deal with problem solving, reasoning, communication, connections, estimation, hands-on activities, cooperative learning, the use of technology, and ongoing assessment. Part of each session is devoted to discussing and reflecting on teachers' perceptions of the after-school laboratories—perceptions based on their observation and participation, including *what* was taught, *how* it was taught, and the *effects on students*. In the workshops, project staff model the techniques and approaches we hope the teachers will use in their work with students. We support NCTM's (1991) assumption that "teachers are influenced by the teaching they see and experience" (p. 124). Teachers' knowledge of mathematics content is strengthened through many of the same kinds of experiences that they later will provide for their students.

The after-school laboratories are highly valued by the participating teachers for several reasons. For example, it gives them "a chance to try ideas out, but with a smaller group of children" (class sizes in California are usually in the thirties, whereas the after-school sessions are limited to twenty children). These tryouts give "a better idea of particular revisions that might be needed for a given class," and the smaller class enables the teachers to "develop a better perception of students' reactions and the needed prerequisites." In particular, the lab setting seems to work much better than just having a sample lesson or doing an activity in a workshop, which might or might not be tried in one's classroom. Combining workshops with opportunities to explore and experiment with new topics and methods in the relatively risk-free environment of the after-school laboratories maximizes the potential for lasting teacher change.

EFFECTS OF THE PROJECT

The project has had significant impact on preservice teachers. Ten preservice teachers who were project participants have entered the highly selective teaching credential program. Their mathematics methods instructors have commented that these students are among the most outstanding students in their classes, both in their understanding of mathematics content as well as in their teaching effectiveness and enthusiasm for teaching. One of these preservice teachers was selected from approximately 200 student-teachers as the outstanding elementary level student-teacher at our university for spring 1991.

The preservice teachers involved in the SDMEP believed that their involvement had benefited them in many ways. They believed that they had a better idea of what it is like to be a teacher and found more relevance in their university coursework. Other preservice teachers commented that the experience of working in the SDMEP helped them to become more aware of what it was like to teach learners of various abilities. The following quotes were made by participating preservice teachers:

- "All students can learn math—just each student needs to be taught in their own way."

- "Every child has the ability to learn math, as long as it is made interesting for them and they can see its use in the world."
- "There is no 'bad student.' Every child can learn—and will learn with the proper guidance."

The San Diego Mathematics Enhancement Project, with the ultimate goal of helping *all students* achieve their full potential in mathematics, has demonstrated that a program of teacher enhancement that involves in-service teachers, preservice teachers, students, and parents in an after-school mathematics enrichment program can have a powerful impact on all parties involved.

REFERENCES

California State Department of Education. *Mathematics Curriculum Framework for California Public Schools, Grades Kindergarten through 12*. Sacramento, Calif.: The Department, 1992.

National Council of Teachers of Mathematics. *Curriculum and Evaluation Standards for School Mathematics*. Reston, Va.: The Council, 1989.

———. *Professional Standards for Teaching Mathematics*. Reston, Va.: The Council, 1991.

National Research Council. *Everybody Counts: A Report to the Nation on the Future of Mathematics Education*. Washington, D.C.: National Academy Press, 1989.

Quality Education for Minorities Project. *Education That Works: An Action Plan for the Education of Minorities*. Cambridge, Mass.: Massachusetts Institute of Technology, 1990.

19

Partnership in Mathematics Education: The Evolution of a Professional Development School

Frances R. Curcio
with
Rossana Perez
Barbara Stewart

L EADERS in government, business, and industry are demanding that schools prepare responsible citizens and competent workers capable of meeting the global and national economic challenges that will be encountered in the twenty-first century. The mathematics education community is responding to these demands by recommending ways to help students develop mathematical power (National Council of Teachers of Mathematics 1989, 1991). Providing students with opportunities to develop mathematical power requires that we make changes in the traditional computation-dominated mathematics curriculum and the "show and tell" methods of instruction. One of the difficulties of effecting change in our schools is that teachers continue to teach the way they were taught (Goodlad 1990). Furthermore, the limited knowledge base of many elementary and middle school teachers and their negative attitudes toward mathematics compound this difficulty.

Collaborative efforts on the part of public schools and colleges can contribute to overcoming some of the obstacles to effecting change. Guidelines for establishing professional development schools have been presented by the Holmes Group (1990). Unlike the laboratory and portal schools that were popular in previous decades (Winitzky, Stoddart, and O'Keefe 1992), the professional development school represents a true partnership between a public school and a college. According to the Holmes Group (1990), a professional development school is much more than a combination of a laboratory school for conducting research projects, a demonstration school to model exemplary teaching, and a clinical site for preparing student and intern teachers. A true professional development school must be designed "for the development of novice professionals, for continuing development of experienced professionals, and *for the research and development of the teaching profession*" (p. 1).

This article describes a unique urban collaborative between the New York City Board of Education and Queens College of the City University of New York. Examples in mathematics education will be used to illustrate the features of the collaborative that are contributing to the evolution of a professional development school.

THE PARTNERSHIP

Since 1979, the New York City Board of Education and Queens College of the City University of New York have enjoyed a thriving partnership (Trubowitz 1986; Trubowitz et al. 1984). We believe that this partnership, which has been centered at the Louis Armstrong Middle School, an urban, New York City public school located in East Elmhurst, Queens, offers a viable model for improving preservice and in-service teacher education in mathematics and that it can offer a rich setting to study the teaching and learning of mathematics.

About the School

There are approximately 1300 students enrolled in grades 5–8 at the Louis Armstrong Middle School. As a magnet school and the only New York City middle school that is not under the auspices of one of the thirty-two local community school boards, students apply to attend. Students are selected to reflect the ethnic and racial composition of the Borough of Queens, stated in the *New York Times* as having the most ethnically diverse communities in the world (Sontag 1992). The proportion of students accepted must also support the mandated varied ability levels determined by standardized tests in reading and mathematics. As a result, 25 percent of the students are above average, 50 percent are average, and 25 percent are below average. The philosophy for organizing students for instruction is supportive of heterogeneous grouping. Some of the many challenges encountered by the middle school teachers in this urban setting include looking for ways to motivate at-risk students, implementing alternative and innovative teaching methods to meet the needs of a multicultural population, and creating a supportive learning environment for children, many of whom live in single-parent homes.

The Nature of the Partnership

Any partnership between a school and a college needs to consider their very different cultures and environments. At times their cultures are in conflict, and the differences need to be negotiated. For example, the two institutions differ in their responsibilities to state and local mandates, the expectations of their constituencies, their means of governance, their sense of time, and their reward systems.

The school is required to meet state and local curriculum mandates that are controlled by scheduled standardized tests; the college has no such curriculum constraints. The school experiences pressure from parent-interest groups and must answer directly to them; the college also experiences pressure from the community but tends to act more independently. The school functions as part of a bureaucracy whereby teachers are held accountable for fulfilling bureaucratic mandates; college professors have more freedom and function collegially. (Traces of bureaucratic constraints in the college's daily operation persist, but they are often ignored by faculty.) The traditional school day is defined by set periods, which are, for the most part, punctuated by the sound of bells; college faculty have less imposing time constraints. Traditional school values include control, student achievement, and cooperation with administrators, all of which influence the teachers' reward system; traditional college values include research and peer-reviewed publications, without which a faculty member would "perish."

A successful partnership not only recognizes the differences in cultures but realizes the value in struggling to overcome problems that may be caused by these differences. Although administrative support for the partnership is essential, both college and school faculties contribute to the ongoing success of the partnership.

A College Professor's Perspective

Queens College professors are a common sight at the Louis Armstrong Middle School (LAMS). During the course of an academic year, many professors spend several days a week working closely with teachers and students. The experience is energizing and enriching. The professors are able to make methods classes relevant to current needs and practices. Although it is very easy to take a "Do as I say" attitude when teaching a methods course, actually modeling a teaching technique in front of student-teachers, graduate interns, and in-service teachers makes the value of the method much more convincing. It also provides a "testing ground" for trying out some new techniques before expounding their virtues.

A college professor recalls how the collaboration in mathematics evolved:

> In the spring of 1986, as an assistant professor completing my first year on the faculty in the School of Education at Queens College, the dean recommended that I visit the Louis Armstrong Middle School (LAMS) to see whether I might be interested in collaborating on a project with some mathematics teachers. I continue to be indebted to her for offering me this suggestion.

> When I started working at LAMS in the fall of 1986, I really did not know how I would fit in. I did not want to impose myself on any of the mathematics teachers, and I certainly did not want anyone to feel threatened by my presence. I was interested in learning *with* and *from* my middle school colleagues, but I was not sure how I could communicate this to them. I found one teacher who expressed an interest in working with me. Our relationship began when I told her that I could bring some manipulative materials from the college. Doing so made me popular among other teachers who were looking for a way to obtain materials. As a result, I was welcomed into other classrooms. For

the most part, I became an extra pair of hands, working more like a student-teacher or an assistant teacher than as a college professor—but I didn't mind. Gradually, after several months I felt comfortable suggesting that we try something a little different—introducing a nontraditional, nonroutine "problem of the week" that we would assign to the students and ask them to work on for a week before discussing it. This proved to be very successful, and the problems found their way into other teachers' classrooms.

A lot has happened to me since those early days—I have learned so much about the daily realities of the classroom, and I have experienced the joys of watching students constructing and discussing mathematics. Working at LAMS has been a highlight for me during my tenure at Queens College.

The School Teachers' Perspective

Although some teachers treat their teaching responsibilities as a "job," many believe that learning how to teach is an ongoing process. Looking for ways to improve teaching is a welcomed challenge, which successful collaboration can help teachers to meet.

Two grade 6 teachers describe how they view the collaboration:

We believe that at least two elements have contributed substantially to our successful school-college partnership: a willingness to work together based on a respect for ideas, and trust built on an openness to discuss problems, difficulties, and misunderstandings. We experience these elements as we continue to collaborate on mathematics curriculum and staff development projects. The willingness to work together has developed because of a genuine interest in learning and sharing that the teachers and the professor bring to the collaboration. The college faculty member has helped us build our confidence by respecting our ideas. We are not afraid to say, "I don't know." We do not feel threatened by the professor because she learns from us just as much as we learn from her. When we are involved in team teaching with the professor, we enjoy the give-and-take attitude that develops between us and among the students. Our questioning techniques have improved, our understanding of mathematics has been strengthened, and our interest in mathematics has blossomed.

We have been collaborating by team teaching, developing alternative curriculum models in mathematics, conducting staff development workshops, and working with preservice teachers since 1986. The projects and activities that have developed result from our mutual concerns, interests, and ideas. Aspects of this "organic collaboration" (Dixon and Ishler 1992), with illustrations in mathematics education, provide an example of an evolving professional development school.

THE EVOLUTION OF A PROFESSIONAL DEVELOPMENT SCHOOL

A by-product of the partnership between the New York City Board of Education and Queens College is the creation of a school that is becoming a professional development school. The Louis Armstrong Middle School supports the development of beginning teachers, provides opportunities for ongoing staff development, and creates a rich setting to study teaching and learning. Examples from mathematics education will be used to illustrate these functions.

Furthermore, many of the six principles on how a professional development school should organize itself, identified by the Holmes Group (1990), are reflected in the daily operation of the Louis Armstrong Middle School. In particular, "(1) teaching and learning for understanding, (2) creating a learning community, (3) teaching and learning for understanding for all children, (4) continuing learning by teachers, teacher educators, and administrators, (5) thoughtful long-term inquiry into teaching and learning, and (6) inventing a new institution" (p. i) will be discussed in reference to the examples presented.

The Development of Beginning Teachers

Many preservice and beginning teachers have a biased view of the teaching and learning of mathematics based on their limited experiences with drill-and-practice worksheets from their own student days in elementary and middle school. Making learning meaningful and teaching for understanding are novelties for most of them. Finding ways to capitalize on the experience of exemplary in-service mathematics teachers and designing a realistic setting for preservice teachers to explore ways of teaching for understanding led to the development of strong ties among the elements of a true learning community. Bringing together in-service and preservice teachers to share their ideas, concerns, and insights became one of the benefits of an after-school mathematics program offered in conjunction with one section of an undergraduate mathematics methods course. (The idea for the After-School Mathematics Program was adapted from Professors James V. Bruni and Helene J. Silverman, Herbert H. Lehman College of the City University of New York.)

Children in grades 4–6 were invited to attend a seven-week, tuition-free program for one hour a week. Under the supervision of three middle school teachers and a college professor, the undergraduates were assigned to work with small, heterogeneous groups of children (i.e., three to five children). The children furnished the undergraduates with experiences that reflect the realities of the urban school. The multicultural population challenged the prospective teachers' creativity in examining teaching and learning styles, developing basic skills, and making mathematics meaningful.

The undergraduates were expected to use games and manipulative materials to teach concepts in middle school mathematics. In preparation for the after-school experience, the undergraduates spent six weeks becoming familiar with nontraditional, exploratory teaching techniques at the college. The undergraduates had the opportunity to experience the fun of teaching mathematics by adapting some of the activities from the *Middle Grades Mathematics Project* (Lappan et al. 1986). The techniques employed support the recommendations of the *Curriculum and Evaluation Standards for School Mathematics* (NCTM 1989).

During the seven-week after-school mathematics program, which was held at the middle school, the teachers assisted and guided the undergraduates as they worked with the children. One difficulty the undergraduates have is eliciting the "big" mathematical ideas implicit in the games and activities they introduce to the children. By joining different groups and playing the games with the undergraduates and the children, the teachers helped the undergraduates formulate questions to elicit the "big" mathematical ideas. For example, a popular game that the undergraduates selected to play with the children was the factor game (Fitzgerald et al. 1986). An objective of this game is for the children to develop "best-move strategies" that are related to examining abundant and deficient numbers. The undergraduates, who have experience playing the game before presenting it to the children, are often not sure of the underlying mathematical concepts involved. Through questioning, the mentor teachers help the undergraduates and the children focus on the number of factors for the numbers. Together, they analyze the factors and discuss the "best-move strategies" without "giving away" any information. The teachers are sensitive to the needs of both the undergraduates and the children. Without embarrassing anyone, they maintain a professional yet friendly and helpful atmosphere.

A by-product of the after-school experience is in-service staff development. The mentor teachers also learn from the undergraduates. Since the teachers are involved in the games, they often find some ideas to introduce in their own classes. For example, several years ago, the game Contig (Broadbent 1975) found its way into a sixth-grade classroom.

After each one-hour session with the children, the mentor teachers join a seminar meeting with the college professor and the undergraduates to discuss their observations, recommendations, and suggestions. At this time, the undergraduates are able to question the teachers about how they plan for instruction, how they deal with disciplinary situations, how they evaluate student progress, how they manage the use of manipulatives, and how they manage learning centers and cooperative learning groups. The rich interaction between the preservice and in-service teachers is beneficial for everyone. As a result of the after-school experience with the teachers, the undergraduates made the following comments:

- "The middle school teachers helped me to understand and appreciate the kinds of knowledge I will need to be an effective teacher."
- "I felt comfortable asking one of the teachers for help. I didn't feel threatened by her because she wasn't giving me a grade. She understood my problem and was eager to offer suggestions."
- "This was a true methods course—the combination of theory and practice was truly rewarding for me!"
- "The after-school mathematics program gave me a chance to see how much fun it is to teach and learn math! I wish we had more time together."

All the children, regardless of their past performance in mathematics, have an equal opportunity to be challenged and to experience mathematics in a meaningful way. The value of the after-school program for the children is evident in their evaluation of the program:

- "Learning math by playing games is really fun!"
- "I really like my tutor. I wish we could play math every day."
- "I wish this program wouldn't end."

Although school teachers have worked with preservice teachers in the capacity of a cooperating teacher, they are experiencing a new role—that of clinical professor. In this capacity, they add a dimension to preservice education that cannot be fulfilled by the professor working alone. Teaching and learning for understanding comes alive for the preservice teachers, the in-service teachers, the teacher educator, and the children involved in the program. Furthermore, by engaging in discussions with in-service teachers about the mathematics, the methods of teaching, and the children's progress, the value of thoughtful inquiry into teaching and learning is instilled in the preservice teachers.

Ongoing Staff Development

The demands and challenges of working in an urban middle school may contribute to teacher burnout. Providing opportunities for professional growth and recognizing exemplary teachers for their contributions help to build a strong learning community.

To provide for ongoing staff development in mathematics, several teachers collaborating with a professor submitted an application to participate in the Middle Grades Mathematics Project Professional Development Teams Workshop at Michigan State University. Our team was selected to participate in the project during the summers of 1989 and 1990. The teachers have become adept in conducting staff development workshops attended by their colleagues, administrators, and teacher educators. They have also been invited to conduct workshops at local professional conferences. Planning for such presentations includes a discussion of both affective and cognitive objectives.

Our affective objectives focus on selecting activities and approaches that will motivate and excite teachers to try some new ideas with their students. Our cognitive objectives focus on mathematics content not only for the teachers' students but for the teachers themselves. The materials and activities in the *Middle Grades Mathematics Project* (Lappan et al. 1986) have helped us accomplish our goals. Although we realize that changing teacher behavior does not occur overnight, we give on-site support to any teachers who want to observe our classes as they use the materials or who want us to visit their classes to offer suggestions.

In 1990, our middle school team was invited by the directors of the Connected Mathematics Project at Michigan State University to establish one of six national professional development centers for middle school mathematics. With the support of the administrators at Queens College and the Louis Armstrong Middle School, the team accepted the invitation. Although the team originally consisted of four people, we have increased our core by offering workshops for teachers interested in assisting us in establishing the professional development center. The enthusiasm and excitement about the project have given the mathematics teachers a sense of professionalism and pride. Participating in a national network produces a setting in which we can continue to learn and share our ideas for improving mathematics instruction.

School teachers are assuming the role of on-site staff developers as they become recognized for their expertise. They are involved in designing and conducting workshops based on periodic needs assessments.

The Study of Teaching and Learning

Research and development related to the teaching profession—in particular, to mathematics—is integrally related to the development of novice teachers and the continuing development of experienced professionals. Research questions arise from interacting with children, undergraduates, and experienced teachers. Sometimes, teachers' informal observations lead to the design of formal research studies. Several longitudinal collaborative projects are supplying us with data that will enable us to continue to refine and enhance what we do as professionals. In particular, projects have focused on studying (1) the effects of small-group interaction on the development of problem-solving skills and (2) how students communicate mathematical ideas by representing data in visual displays.

A team of mathematics teachers is currently involved in field-testing an innovative mathematics curriculum developed by the Connected Mathematics Project at Michigan State University. The teachers review materials, test them out with their students, and make suggestions for improving them. Being involved in the development of materials has added a new dimension to the role of the mathematics teachers at the middle school. They are actively involved in shaping and designing materials appropriate for students in a multicultural environment.

Another curriculum development project is the result of a collaboration between a grade 7 teacher and a professor. Unhappy with a traditional grade 7 mathematics curriculum that is dominated by the review of concepts from grades 5 and 6, they designed an experimental grade 7 curriculum in the format of a problem-solving seminar that incorporates the *Challenge of the Unknown* videotape series (Maddux 1986). The problem-solving seminar, designed for heterogeneous groups of seventh graders, incorporates technology, cooperative learning, and reading and writing in the mathematics class. Sessions taking place before and after the viewing of the videotapes include students actively involved in a series of experiments and investigations (Curcio, McNeece, and Rosen 1988).

Innovative and experimental ideas generate naturally as a result of having teachers and professors interacting at the middle school. As a result of these types of activities, teachers are assuming new roles—those of curriculum developer and researcher.

NEXT STEPS IN THE EVOLUTION OF A PROFESSIONAL DEVELOPMENT SCHOOL

Although the Louis Armstrong Middle School has many of the features of a professional development school described by the Holmes Group (1990), the school has a long way to go as it continues to evolve into a new institution. The traditional roles of teacher and college professor are shifting to take on new dimensions. For example, as described previously, teachers have become clinical professors, curriculum developers, researchers, and instructional decision makers. The college professor has become a classroom teacher, the director of an after-school program, and a facilitator, bringing together essential elements of the learning community. As a result of these new roles, expectations and reward systems need to be reexamined. Teachers must no longer be viewed and employed as hall monitors, lunchroom supervisors, and inventory clerks. College faculty must no longer be judged only by the publication of their theoretical papers. We need a new set of standards for professionals operating in this "new" institution.

All the activities described are ongoing, dynamic efforts that require patience, time, and administrative and financial support. Commitment to strengthening the ties between the school and the college must be continually reaffirmed. Teacher educators are expected to be in touch with the reality of the public schools if they are to prepare future teachers effectively. More college faculty need to build personal partnerships with school teachers and work closely with students in the schools. The challenge confronting us is how to entice them to seek out such partnerships.

We need to strengthen how we document what we do in the area of research and development to promote teaching as a profession. Long-range goals and ongoing projects must have clearly defined means of evaluation. How to document progress over time systematically needs to be a top priority.

CONCLUDING REMARKS

A school-college partnership such as the one described in this article takes hard work on the part of everyone involved. It is an important relationship, however—one that should be nurtured and cultivated.

A professional development school is a natural setting in which professors can keep in touch with the reality of the classroom. Exemplary teachers can be recognized and used to supplement methods courses with their own insights, techniques, and on-site expertise. Moreover, in-service professional development is an outcome that generates renewed enthusiasm. Furnishing teachers with challenging opportunities for professional growth will help to reduce the chances of teacher burnout and keep competent teachers in the classroom. Improving preservice and in-service teacher education at Queens College and the Louis Armstrong Middle School was a by-product of the energy and excitement generated by the children, teachers, undergraduates, and the professor all working together.

REFERENCES

Broadbent, Frank. "'Contig': A Game to Practice and Sharpen Skills and Facts in the Four Fundamental Operations." In *Games and Puzzles for Elementary and Middle School Mathematics: Readings from the "Arithmetic Teacher,"* edited by Seaton E. Smith, Jr., and Carl A. Backman. Reston, Va.: National Council of Teachers of Mathematics, 1975.

Curcio, Frances R., J. Lewis McNeece, and Jamie Rosen. "What's the Problem? A Model for Curriculum Innovation in Mathematics." *Connections* 14 (Spring 1988): 8–10.

Dixon, Paul N., and Richard E. Ishler. "Professional Development Schools: Stages in Collaboration." *Journal of Teacher Education* 43 (January-February 1992): 28–34.

Fitzgerald, William, Mary Jean Winter, Glenda Lappan, and Elizabeth Phillips. *Factors and Multiples.* Menlo Park, Calif.: Addison-Wesley Publishing Co., 1986.

Goodlad, John O. *Teachers for Our Nation's Schools.* San Francisco: Jossey-Bass Publishers, 1990.

Holmes Group. *Tomorrow's Schools: Principles for the Design of Professional Development Schools.* (ERIC Document Reproduction Service No. ED 328 533.) East Lansing, Mich.: The Group, 1990.

Lappan, Glenda, William Fitzgerald, Elizabeth Phillips, Janet Shroyer, and Mary Jean Winter. *Middle Grades Mathematics Project.* 5 vols. Menlo Park, Calif.: Addison-Wesley Publishing Co., 1986.

Maddux, Hilary C., ed. *The Challenge of the Unknown: Teaching Guide.* New York: W. W. Norton & Co., 1986.

National Council of Teachers of Mathematics. *Curriculum and Evaluation Standards for School Mathematics.* Reston, Va.: The Council, 1989.

———. *Professional Standards for Teaching Mathematics.* Reston, Va.: The Council, 1991.

Sontag, Deborah. "New York City Tops Nation in Immigration." *New York Times,* 1 July 1992, pp. B1, B3.

Trubowitz, Sidney. "Stages in the Development of School-College Collaboration." *Educational Leadership* 43 (February 1986): 18–21.

Trubowitz, Sidney, et al. *When a College Works with a Public School: A Case Study of School-College Collaboration.* Boston: Institute for Responsive Education, 1984.

Winitzky, Nancy, Trish Stoddart, and Patti O'Keefe. "Great Expectations: Emergent Professional Development Schools." *Journal of Teacher Education* 43 (January-February 1992): 3–18.

20

Part 3

Empowering K–12 Teachers for Leadership: A Districtwide Strategy for Change

Peggy A. House

"WHAT would you do," the questioner asked, "to make the NCTM *Standards* become a reality instead of just another report in the file cabinet?" The question was right on the money. The *Curriculum and Evaluation Standards for School Mathematics* (National Council of Teachers of Mathematics [NCTM] 1989) had just been published, an enormous accomplishment in its own right. But the real challenges for mathematics educators lay ahead.

During the previous year, when the *Standards* was being circulated in draft version, I had had the opportunity to study and discuss the document with more than one hundred mathematics teachers of all grade levels, K–16. Predictably, their initial reactions ran the gamut from unrestrained endorsement to cautious realism to extreme skepticism and even outright rejection. But as they began to formulate their responses to the proposed *Standards,* a curious pattern emerged, even among the most optimistic. Each, it seems, perceived that the major obstacles to the implementation of the Standards were all the other teachers in his or her respective school.

Convinced that the success or failure of mathematics education reform lay in the hands of individual teachers working cooperatively, I described my plan for the inquisitor. If I had the proverbial three wishes, I told him, I would wish for the following things: first, a school district sincerely open to the possibility of change—not necessarily already committed to the Standards but willing to take the risk of examining current practices and trying new things; second, administrative leadership and district support for the efforts to be undertaken; and third, a core of K–12 teachers who have the courage to become leaders in their own district. Given these three ingredients, significant changes would be within reach.

FACILITATING CHANGE IN EDUCATION

Changes in educational practices do not come quickly or easily; lasting changes are rarely imposed from without. Indeed, the research on educational change has been clear in describing certain strategies that lead to successful long-term instructional change. In particular, the following four characteristics are common to such approaches: (1) the local school site is the locus of change, (2) teachers are involved in the innovation-related decision making, (3) the school adopts a problem-solving orientation, and (4) the change process focuses on problems that teachers consider important to their day-to-day routines (Williams and Cannings 1981, p. 22).

The National Research Council also recognized the essential role of teachers when it called for a national strategy to revitalize mathematics education that would employ an "augmented grass-roots" model of curricular development to harness the power of a centralized system with the flexibility and initiative of the decentralized U.S. tradition (National Research Council 1989, p. 90). In mathematics education, the coordinated national leadership has been provided by NCTM through the *Curriculum and Evaluation Standards for School Mathematics* (1989) and the *Professional Standards for Teaching Mathematics* (1991), both of which were developed by national teams of teachers, teacher educators, supervisors, and others concerned with mathematics education. But NCTM deliberately crafted the *Standards* documents to present a set of goals and a vision of mathematics teaching "without a prescription for achieving them" (NCTM 1989, p. 251). Thus, the other component, local ownership leading to creative implementation, is the professional responsibility of every mathematics teacher. The project described in this paper represents one school district's response to the challenge of implemention.

ACCEPTING THE CHALLENGE

The opportunity to fulfill the three wishes described above came shortly after the initial question was posed. The school district of White Bear Lake, Minnesota, a middle-class suburb of the Twin Cities, responded favorably to an inquiry about their interest. The district, which serves approximately five thousand elementary school pupils in ten K–6 schools and an additional thirty-five hundred secondary school pupils in one junior (grades 7–8) and two senior (grades 9–10 and 11–12, respectively) high schools, had a reputation for supporting innovations in the mathematics program, although none as extensive as this one had yet been undertaken.

With support from the Dwight D. Eisenhower Mathematics and Science Education Program, a two-year project of professional growth and instructional change was designed with the following goals:

- To involve teachers and administrators from all levels of the K–12 system in a process to identify specific needs for improvement in the mathematics curriculum

- To empower a leadership team, with membership from each of the district's elementary, middle, and secondary schools, to model

desired classroom teaching strategies consistent with the vision outlined in the *Curriculum and Evaluation Standards* and with the state
and local goals for the mathematics curriculum

- To involve the leadership team in the production of demonstration
  lessons that would become the basis of in-service programs in the
  district's schools

- To develop a program of district in-service training in mathematics to be
  conducted at the building level by the members of the leadership team

- To evaluate the activities undertaken during the project and to disseminate both the process and the products of the program to other
  school districts

Forty-six participants formed a leadership team consisting of at least two classroom teachers from each of the district's ten elementary schools; five mathematics teachers from the junior high school; four mathematics teachers from
each of the two senior high school buildings; district personnel with responsibilities for special education, remedial education, gifted education, media and
resources, staff development, and administration; and one classroom teacher
from a parochial elementary school in the district. All the team members were
volunteers selected by their building principals and a five-member steering committee consisting of one university mathematics educator; one elementary school
principal, who also chaired the district's elementary school mathematics committee; the junior and senior high school assistant principals who were cochairs of
the district's secondary school mathematics curriculum committee; and one
junior high school mathematics teacher. Although administrative membership on
the steering committee was disproportionately high, it was deemed important in
demonstrating the district's support for the project and the priority assigned to it.

UNDERLYING PREMISES

Lasting changes in mathematics classrooms result only when teachers confront their beliefs about what mathematics is, what it means to do mathematics,
and how mathematics is taught and learned. Examining these beliefs, in turn,
demands ownership of the need to change, consensus on the desired direction
for change, and acceptance of the change process. Too many attempts to change
education have failed because they were imposed in a top-down manner without
the significant involvement of the teachers who must in the end be the primary
agents of change. Consequently, the project described here was built on the following two premises: First, teachers must be centrally involved in identifying
and planning the specific changes to be undertaken. This involvement requires
ample time and opportunity for communication within and across grade levels,
and it demands an environment of mutual trust and respect that supports teachers in taking risks. Second, the undertaking would be a unified K–12 initiative

that seeks to engender common goals, common beliefs about mathematics, and common aspirations for the mathematics program. Thus, the Standards were to be viewed as a seamless whole, not a patchwork of grade-level reforms, and all participants would be expected to read the entire *Curriculum and Evaluation Standards* and to participate in all project activities, regardless of the focus of the lesson or the teaching assignment of the teacher.

From the onset, the steering committee agreed on several strategies that would guide the undertaking. In particular, care would be taken to avoid any criticism of teachers' past teaching practices. Nothing would be gained by telling teachers not to teach X or that they had been teaching Y improperly; to do so would only establish a confrontational environment in which the participants felt threatened and defensive. Neither would we attempt to overhaul the entire curriculum or make sweeping instructional changes in the immediate future. Instead, the project would focus positively on four major goals:

- To build number sense and an awareness of the mathematical nature of the world around us
- To emphasize mathematical reasoning and the ability to represent mathematical concepts in concrete contexts
- To foster mathematical communication and the ability to explain one's reasoning
- To strengthen the mathematical self-confidence and self-reliance of both the teachers and their students

None of these goals proved threatening to the participants, and all the participants agreed that they were important but underachieved in typical classrooms.

PROJECT OVERVIEW

The two-year project evolved through several identifiable but overlapping stages, as follows:

Phase 1: Awareness and Familiarity
(October through December, Year 1)

Each of four all-day workshops scheduled at approximately three-week intervals included a formal presentation and hands-on activities led by resource persons familiar with the NCTM *Curriculum and Evaluation Standards* and the state's learner outcomes. Resource persons included classroom teachers and a mathematics coordinator from neighboring districts, the mathematics specialist from the state department of education, and university educators. The first workshop presented an overview of mathematics education worldwide and of the NCTM Standards in relation to state and local goals for mathematics instruction; the next three workshops examined the K–4, 5–8, and 9–12 Curriculum

Standards in greater depth. The discussion emphasized the unity of the K–12 Standards and the importance of all teachers' being familiar with the goals and activities of the mathematics program at all levels. All the participants took part in all the activities and discussions, regardless of the teaching assignment of the participant or the level of mathematics under discussion. Each day began with a presentation that included examples of lessons and activities that promote the goals of the *Curriculum and Evaluation Standards* and the *Professional Teaching Standards,* and each workshop included time for the participants to meet in smaller working groups to assess the strengths and needs of the mathematics program at each grade level and to plan model lessons that exemplify desired teaching strategies and learner outcomes.

Phase 2: Experimentation
(November through January, Year 1)

The second stage emphasized implementing desired strategies in the classrooms of the participants. Working in grade-level groups (early primary, late primary, intermediate and junior high school, and high school), the teachers planned activity lessons that each one taught in at least one class. The lessons were modeled on activities done in the workshops, and the results of the lessons, including samples of students' work, were discussed and evaluated at the next session. In subsequent sessions, teachers planned and taught longer units and cooperatively developed plans for adapting the curriculum that they would be teaching during the coming months.

Phase 3: Integration
(January through May, Year 1)

During the second half of the first year, the teachers strove to infuse their regular curriculum with strategies that emphasized the reasoning and communication goals described above. They taught lessons that they had developed cooperatively, and each arranged to have one or more classes videotaped. At the final in-service day of the first year, the teachers met in small multigrade groups to review and discuss the vidoetapes of one another's classes and to begin to formulate a district plan based on identified needs in the individual schools. The video tapes brought by the teachers were collected for review and editing so that selected portions could be used as demonstration lessons in future in-service activities with the rest of the district staff.

Phase 4: Assessment
(Ongoing from May of Year 1)

Data collected from the members of the leadership team were organized over the summer and shared with the participants in September. During an initial all-day workshop in the fall, the participants discussed the data from the project, evaluated the progress to date, and made plans for the next steps. Attention was given to coordinating the work of the leadership team with other district initiatives, including textbook adoptions and a pending accreditation review.

Phase 5: Implementation
(October through May, Year 2)

The second year of the project included three all-day workshops (during October, November, and January) during which the leadership team addressed the question of how to share the project with other teachers in the district. During the second year, four teachers from each of the ten elementary schools and the remaining eleven secondary school mathematics teachers were added to the project. This expanded team of twenty-seven grades K–4 teachers, eighteen grades 5–8 teachers, and six grades 9–12 teachers received two full days of in-service training conducted by members of the leadership team. To help the team members prepare for their in-service presentations, the fall workshop days for the leadership team included presentations by a district administrator and steering committee member on characteristics of effective schools and by two district staff development specialists, both members of the leadership team, on strategies for planning effective in-service programs and on working with adult learners. The selection of school district personnel to conduct these sessions was deliberate and designed to reinforce the message that the expertise within the district can and should be tapped. Later, teachers on the leadership team would similarly become resources for their peers.

A first round of all-day in-service workshops for the expanded team was held in the fall with follow-up workshops in January. In the fall, separate workshops were held for grades K–4 teachers, for grades 5–8 teachers, and for grades 9–12 teachers. In January, the sessions were held for five of the K–6 elementary schools, for the other five K–6 schools, and for secondary school (grades 7–12) teachers. This arrangement was decided by the teachers as a way of bridging the elementary school–secondary school gap by aligning the junior high school teachers with both the middle grades teachers and the senior high school faculty, and participants were particularly pleased with the arrangement.

The workshops for the expanded group, which were planned by the entire leadership team and conducted by ten of its members, stressed developing understanding of the Standards, examining one's own beliefs about mathematics and the way it should be taught, modeling lessons that exemplify the principles of the *Standards* documents, engaging participants in mathematics activities, and assisting participants in implementing similar activities in their own classrooms. In a condensed form, the workshops mirrored the development of the first-year program in which the leaders had participated.

Other activities during Year 2, conducted by various members of the leadership team, included (*a*) a series of monthly one-hour mini–in-service meetings in each elementary school for the faculty who were not on the expanded team; (*b*) planning outreach activities to inform district administrators, parents, and others of proposed changes in the mathematics program; (*c*) coordinating project activities with the district's accreditation review and with textbook-selection activities; (*d*) editing a set of demonstration lessons taught by project participants for use in in-service presentations within the district; (*e*) developing model lessons and supporting materials for use by district teachers wishing to try some of the recommended strategies and activities; and (*f*) preparing presentations to be made by several of the participants at the annual meeting of the state's NCTM affiliate.

A final meeting of the leadership team was held in May of the second year. The focus of the day was on the newly released *Professional Standards for Teaching Mathematics* (NCTM 1991). Teachers reflected on the four themes of the *Professional Teaching Standards*—tasks, discourse, environment, and analysis—and on the ways in which their mathematics teaching had changed during the two years of the project.

Phase 6: Routine Use
(Ongoing from Year 3)

After completing the activities described in Phase 5 above, members of the leadership team, together with some members of the expanded team, met in October of the following year to propose strategies for the long-term implementation of the changes recommended and developed during the first two years. Continued monitoring and improvement of the mathematics program remain districtwide priorities now and into the future.

One measure of the success of the program can be found in the attention it received from neighboring school districts. In the year following the conclusion of the formal project activities, two neighboring school districts launched a joint program supported by Eisenhower funding for their K–8 mathematics teachers. Eight members of the original leadership team agreed to serve as teacher-mentors for the two new districts. In that capacity, they conducted four all-day in-service workshops for peers in the new districts. In addition, sixty-five teachers from the new districts made full-day visits to schools in the original project to observe mathematics teaching there, and teachers from the original leadership team made classroom visits to the neighboring teachers. Thus the program had a ripple effect through the professional interaction and mentor relationship established among the three school districts.

DISCUSSION

Several features contributed to the success of the program. Among these were attention to beliefs and expectations, opportunities for networking and interaction, duration and continuity, and administrative support.

Beliefs and Expectations

Any attempt to modify teachers' behaviors without at the same time modifying their beliefs and expectations is likely to produce transient results at best. The mathematics education reforms envisioned in the *Standards* documents are predicated on a conceptualization of mathematics as a dynamic human activity and on the belief that all persons are capable of developing mathematical power. Yet numerous studies have concluded that in American society, the prevailing image of mathematics is that of a collection of facts and rules and that the prevailing belief about mathematical ability is that one is either born with mathematical genes or doomed to a life of mechanical calculations with, at best, limited understanding. Unfortunately, large numbers of teachers, consciously or unconsciously, also espouse such beliefs.

With these misimpressions in mind, considerable time and attention were given to the matter of beliefs and expectations. On the first in-service day, teachers responded to a series of sentence-completion stimuli:

• Mathematics is....
• We teach mathematics because....
• The most important reason for knowing mathematics is....
• In order to learn mathematics, a student must....
• Compared to other school subjects, mathematics is....
• In my mathematics classes, students spend most of their time....
• The environment in my mathematics classes can best be described as....
• Most students think mathematics teachers are....

Teachers also predicted how they thought their students would respond to similar questions. In groups, they formulated a composite picture of what seemed to be their own prevailing images of—

• the nature of mathematics;
• the way a person learns mathematics;
• the role of the mathematics teacher;
• the role of the mathematics student;
• the characteristics of a person who is "good at mathematics."

Also, grade-level teams each developed a plan for surveying the beliefs of their own students, of other teachers in their schools, and of parents and other adults. Each teacher was charged with the task of "taking the mathematical pulse" of his or her own school community and of summarizing the findings about the five dimensions listed above. Further, each was to characterize the vision of the *Standards* documents on the same five dimensions and to compare that picture with the prevailing district beliefs.

At the next meeting, the teachers shared their summaries as well as copies of individual pupils' and adults' comments. Individually and collectively, the results reflected the status quo: mathematics was commonly thought to be about memorizing and applying rules; the teacher's role was described in terms of the dispenser-of-information and trustee-of-right-answers metaphors; students were expected to listen and pay attention and absorb what they were told and shown; and persons who are "good at mathematics" were stereotyped as conforming with expectations for classroom behavior and completing assignments, applying calculation algorithms rapidly and accurately, and "having a mathematical mind." The contrast with the vision in the *Standards* documents was striking.

Overall, teachers were appalled at the results. One described it as "the most depressing assignment I ever had." Asked why, she replied, "Because I knew *other* people believed those things, but I never imagined that *my students* did!"

The exercise was significant in raising the consciousness of the participants. It served as an impetus to examine their classroom behaviors, both verbal and nonverbal, in the light of the expectations that teachers set for both themselves and their students. Over the duration of the project, the teachers became more acutely aware of the attitudes they conveyed, of teacher and student behaviors that revealed underlying beliefs, and of the importance of building positive images and expectations about mathematics. It is significant that one year later, when the leadership team identified goals for the in-service program that they were developing, their list revealed beliefs and goals strongly in conformity with the *Standards* documents. Their composite answer to the question "What do we want to teach our peers?" was as follows:

- Communicate our excitement
- Show them reasons why we need to change
- Familiarize them with the Standards
- Help them examine their beliefs
- Let them share their concerns
- Examine what needs to change and what should remain the same
- Give concrete examples and demonstration lessons that model the way we want to teach
- Guide them through a process of modifying a traditional lesson
- Introduce them to problem solving, manipulatives, existing resource materials, uses of technology, and strategies for small-group activities
- Examine issues related to evaluation
- Explore how to integrate mathematics with other subjects
- Discuss new roles for teachers and students
- Relate what we're doing to our district and state goals
- Assure them that it's okay to try new strategies and that they don't have to "teach the book"

- Bolster their confidence
- Raise their expectations
- Validate what we're all trying to do

The most seasoned mathematics supervisor would be hard pressed to construct a more comprehensive or insightful list of goals.

Networking and Interaction

A major strength of the program was almost universally identified as being its involvement of both elementary and secondary school teachers, strongly supported by the district, in an effort to implement a grass-roots approach to curricular and instructional change. District officials described the program as one of the most successful efforts to bring together elementary and secondary school teachers in a common effort. The teachers on the leadership team repeatedly emphasized the strengths of the project:

- The interaction with professional peers
- The professional exchange of ideas
- The districtwide, K–12 focus
- The richer perspective on mathematics
- The exposure to new materials and resources
- The support for trying new approaches in individual classrooms
- The validation of individuals' beliefs and efforts

One way in which these strengths were achieved was through allotting time in every workshop for teachers to interact in discussion groups, sometimes in cross-grade groups and sometimes in more homogeneous assemblies. What developed were a new level of mutual respect and an acceptance of one another as partners on the same mathematical team. The secondary school teachers who viewed videotapes of elementary school mathematics lessons were struck by the energy of the elementary school teachers. One high school teacher admitted to being exhausted just from watching the elementary school videotapes; he could not imagine how the teachers maintained that level of activity on a regular basis. The elementary teachers, in contrast, stopped describing themselves as, for instance, "only a second-grade teacher," and junior high school teachers developed strong bonds with the fifth- and sixth-grade teachers as well as with the secondary school staff. Large numbers of elementary school teachers chose to attend state mathematics conferences, and within one year, several of them were making presentations at state and regional mathematics education meetings.

Duration and Continuity

The foregoing accomplishments would not have been possible were it not for the extended duration of the project. Unlike many in-service programs that too often are stand-alone, one-time events, this program had the advantage of continuity over a two-year period. Consequently, participants had the opportunity to evolve in their beliefs and expectations; to develop linkages with colleagues; to try new strategies in their classrooms and reflect on the outcomes; to receive feedback, new ideas, and encouragement from colleagues; and to develop their self-confidence and sense of mathematical power. Workshops were full-day events held during the school day; Eisenhower Program funds were used primarily to support the substitute teachers who made this arrangement possible. The time between meetings was sufficient for teachers to try new approaches in their own classrooms in the interim but not so long as to lose continuity. The participants eagerly looked forward to the sessions, and attendance was high. None of the leaders dropped out of the project.

At the conclusion of the project, the participants expressed their determination to keep alive the positive features of the previous two years. All the members of the district's elementary school mathematics committee were on the leadership team; they became proactive in integrating the project's outcomes into the district's mathematics curriculum, textbook selection, and instructional program. Further, the bridges built between the elementary and secondary schools continue to be maintained; the junior high school mathematics teachers' ongoing interactions with both elementary and senior high school mathematics teachers are particularly instrumental in that effort. The participants were realistic in identifying the following as ingredients needed to sustain the momentum of the project:

- Adequate preparation time
- Financial support for in-service training and materials
- Coordination across all levels
- In-service opportunities for all teachers in the district
- Public acceptance of, and support for, the district's efforts
- Continued evaluation and reporting

Administrative Support

The cooperative arrangement with the school district was positive from the beginning. The interest and enthusiastic leadership demonstrated by the administrators on the steering committee were critical factors in the success of the project; the district's chief administrators and school board gave their unqualified support and encouragement. Meetings of the steering committee were a priority for all members, and their attitude of shared ownership of the success of the project was clearly evident. Not only did the three principals attend all the in-service workshops, but they also took an active role in each one. At every workshop, they led discussions, worked in large and small groups with the participants, and made presentations. Their enthusiastic support for the project and their strong affirmation of the participants contributed significantly to the enthusiasm and high morale of the participants and to the lack of attrition among the volunteers in the project.

CONCLUSION

Lovitt et al. (1990, p. 234) described models of successful in-service programs and concluded that

> although the models may appear different in organizational detail, they are underpinned by the following set of key features identified by research and our experience. To be effective, a professional-development program should—
> * address issues of concern recognized by the teachers themselves;
> * be as close as possible to the teachers' working environment;
> * take place over an extended period of time;
> * have the support of both teachers and the school administration;
> * provide opportunities for reflection and feedback;
> * enable participating teachers to feel a substantial degree of ownership;
> * involve a conscious commitment on the part of the teacher;
> * involve groups of teachers rather than individuals from a school;
> * use the services of a consultant or critical friend.

The project described in this paper had all the aforementioned features. The teachers had individual and collective ownership of the issues and exercised grass-roots leadership in working out the solutions; the administrators were strong champions of the teachers; and a university mathematics educator served as the critical friend.

The eagerness with which teachers not on the leadership team responded to opportunities to learn more about the activities of the project attested to the positive messages that the participants projected to their colleagues. The fact that participants from the project were sought out as speakers for professional meetings and as mentors for teachers in neighboring school districts gave evidence of the respect that other professionals had for the accomplishments of the teachers. The "augmented grass-roots" model (National Research Council 1989, p. 90) for revitalizing mathematics education by empowering teachers to become effective leaders is, indeed, an attainable goal.

REFERENCES

Lovitt, Charles, Max Stephens, Doug Clarke, and Thomas A. Romberg. "Mathematics Teachers Reconceptualizing Their Roles." In *Teaching and Learning Mathematics in the 1990s, 1990 Yearbook of the National Council of Teachers of Mathematics,* edited by Thomas J. Cooney, pp. 229–36. Reston, Va.: The Council, 1990.

National Council of Teachers of Mathematics. *Curriculum and Evaluation Standards for School Mathematics.* Reston, Va.: The Council, 1989.

———. *Professional Standards for Teaching Mathematics.* Reston, Va.: The Council, 1991.

National Research Council, Mathematical Sciences Education Board. *Everybody Counts: A Report to the Nation on the Future of Mathematics Education.* Washington, D.C.: National Academy Press, 1989.

Williams, Richard C., and Terence R. Cannings. "The Dilemma of American Educational Reform." In *Changing School Mathematics: A Responsive Process,* edited by Jack Price and J. D. Gawronski, pp. 13–23. Reston, Va.: National Council of Teachers of Mathematics, 1981.

21

An Agent for Change: The Woodrow Wilson National Fellowship Foundation

Catherine Anne Wick
Susanne K. Westegaard
Carolyn Q. Wilson

THE National Leadership Program for Teachers (NLPT) of the Woodrow Wilson National Fellowship Foundation has been bringing about significant change in science, history, and mathematics education since the first four-week chemistry institute in 1982. Although this article is about the mathematics component of the program, much of what is written here applies to the NLPT in the other disciplines as well. There are three components to the Foundation's Leadership Program for Teachers. The *four-week summer leadership institutes* bring fifty expert mathematics teachers to the campus of Princeton University to work with master teachers and college and university faculty on mathematics content and pedagogy. Two kinds of outreach occur as a direct consequence of the four-week institutes. Teams of teachers made up of participants from the national institutes conduct *week-long institutes* for teachers around the country, beginning the summer after their work at Princeton and continuing year after year. Another kind of outreach involves *one-time local and regional projects* initiated by national institute participants. Each of these aspects of the NLPT will be discussed in more detail.

FOUR-WEEK SUMMER INSTITUTES

Participating in a four-week Woodrow Wilson institute is a life-changing experience. Individual teachers who are successful in their classrooms and committed

Both authors are 1989 Woodrow Wilson Master Teachers. Westegaard is a member of an algebra TORCH team. Wick was assistant academic director for the four-week mathematics institutes in 1991, 1992, and 1993.

to reform in school mathematics are brought together in a setting where their knowledge of teaching is valued. Many are accustomed to being the single voice for change in a school or a district. At Princeton, they live together in a dormitory, eat together, learn together, work together, and discover the value of not being alone on the cutting edge. They are encouraged to expand their vision of themselves and of their profession. At the end of the program, they take away with them the confidence to take risks and the understanding that they are part of a powerful network of teacher-leaders. The impact of a leadership institute is not just for September but also for the longer term.

> "The network of good teachers who are only a phone call away, the exciting material to be shared, the increased confidence that what we do within our own classrooms is valuable and needs to be shared, the delight in learning new topics as we try to share them with students, the pride in making a dent in the system from within—these are some of the elements that the program has encouraged."

The core of the NCTM *Professional Teaching Standards* lies in the beliefs that "teachers are key figures in changing the ways in which mathematics is taught and learned in school" and "such changes require that teachers have long-term support and adequate resources" (NCTM 1991, p. 2). This line of thought is not new. The Woodrow Wilson National Fellowship Foundation, through its National Leadership Program for Mathematics Teachers, has been actively espousing these beliefs, backed with both funds and programs, since 1984.

The NLPT recognized that teachers are professionals. People who have been successfully applying the study of teaching as a skill, if not an art form, are people who have an intuitive sense of what will work in a classroom full of adolescents and what will not. These master teachers can help close the gap between what educational researchers and policy makers believe should be done in the classroom and what actually works. Participants, or "Woodies," enter the Princeton experience as active professional educators who have demonstrated their interest and skill in leadership. They return to their school districts empowered to extend their involvement in improving school mathematics.

> "The institute gave me more new ideas than I thought possible, the desire to incorporate them, and the belief that I can bring about change in my school, my district, and beyond."

The names of Woodrow Wilson Master Teachers appear prominently as presenters at regional and national conferences, journal and yearbook contributors, textbook and resource material authors, software developers, and members of NCTM's and its affiliated groups' boards of directors and and committees. Since the first Presidential Awards for Excellence in Science and Mathematics Teaching, approximately 20 percent of the awardees have been Woodrow Wilson Master Teachers. Their influence is felt among their colleagues, within their communities, and throughout the nation.

> "The institute made me take a serious look at my curriculum, the changes I can make, and how I can influence others to make those same changes."

The Woodrow Wilson Foundation is an administrative organization that bases its policies on research and raises funds to support its programs. The Pew Charitable Trusts and the National Science Foundation are among recent major sponsors. The Woodrow Wilson Foundation organizes the planning for the NLPT through groups of advisors, including members of relevant professional organizations, college and university faculty, teachers, and members of the program staff. This advisory group selects the topics to be addressed and advises on faculty for the four-week institutes held each summer on the campus of Princeton University.

For the institutes of 1984 through 1990 and in 1993, participants were high school teachers from all over the United States. Topics included statistics, geometry, functions, mathematical modeling, human decision making, algebra, and the mathematics of change. In the summers of 1991 and 1992, the focus was on mathematics in the middle school, specifically the topics of shape and uncertainty as developed in *On the Shoulders of Giants* (Steen 1990). The long-term faculty has included expert secondary school teachers and internationally recognized leaders in mathematics content and pedagogy, along with software developers, test writers, researchers, and textbook authors.

> "One of the best aspects of the program was the exposure to the nation's most advanced work on curriculum and curriculum delivery."

The foundation handles all administrative details for the four-week institutes. Before the institute, participants are recruited and selected, planning sessions are held, and arrangements are made for all aspects of the actual institutes, including finances. Participants have their expenses paid and receive a stipend. Staff members oversee such mundane aspects as meal schedules and equipment needs for the fifty participants and the faculty during the institute. Freedom from worry about such details allows participants and faculty to concentrate on what they do best, that is, improving the teaching of mathematics. After the completion of the institute, the foundation's support continues through ongoing encouragement for teachers' outreach activities.

The program of a four-week institute might best be termed a curriculum development laboratory. Through formal presentations, laboratory explorations, and informal work with program directors, visiting faculty, and colleagues, participants examine new concepts, teaching techniques, and technology. Faculty members treat the participants as valued consultants, soliciting opinions about products and ideas from those whom they consider master practitioners.

> "Not only did we listen to them (expert faculty)—for the most part we ate with them and had conversations just like we were *their* colleagues!"

In computer and resource "stuff" laboratories, participants experiment with prototypes for such tools as data-analysis software, spreadsheets, symbol-manipulator software, integer manipulatives, polyhedral models, and graphing calculators. Major blocks of time are provided for teachers to work with materials and ideas, to reflect on how useful they might be in the classroom, and to work in teams to develop short units for such use. These units are published by the foundation, distributed to previous years' "Woodies," and available to anyone, on request, as long as supplies last. (Write to CN 5281, Princeton, NJ 08543-5281.) In addition, the daily schedule includes opportunities for the teachers to present to one another something that works well in their own classrooms.

> "It was so valuable to swap activities and ideas with other teachers who are willing to stretch the limits."

OUTREACH PROGRAMS

The challenge of bringing about change in mathematics classrooms is not completely answered by having great material in a form that is useful to teachers. It must be disseminated to other teachers and implemented in classrooms. Teachers must try new materials, work with them, make them "their own," and improve them until the materials become an integral part of their curricula.

Most professionals learn best from their peers, and teachers are no exception. Teachers are inclined to trust other teachers, those who understand and share the conditions under which they work.

> "A strength of the week-long workshop was having actual class-room teachers do presentations."

The NLPT has made it possible for the master teachers, who have taken part in its program and developed new materials, to spread their work and their influence to other teachers through outreach programs. The common feature of all outreach programs is teachers teaching teachers, furnishing validity for ideas and support for professional growth and classroom change. Woodrow Wilson Master Teachers, teaching other teachers, are effecting change in mathematics classrooms around the United States.

> "Something I particularly liked about the Woodrow Wilson out-reach institute was the "user friendly" teachers—always ready to help or discuss."

Two kinds of outreach follow the four-week summer leadership institutes.

Teacher-Initiated Outreach Projects

One type of NLPT outreach involves small grants—up to about $1000—made available to the participants for local and regional projects. These projects may be as limited as a brief workshop presentation by one master teacher or as extensive as Saturday seminars extending over a period of months and developed and presented by a team. Projects are limited only by the funds available and the imaginations of the master teachers

Outreach projects initiated by the master teachers have often been so successful that additional local and national funds have made possible the continuation of the projects beyond the original design. The teachers have become recognized in their communities and beyond as leaders in professional development and reform activities.

> "Without the Woodrow Wilson summer at Princeton, I would still be an energetic teacher, enjoying my students and trying to do my best. But since the WW summer, I have extended myself to other teachers and I enjoy the new challenges of in-service opportunities."

One-Week Institutes

Another kind of outreach is one-week topical Teacher OutReaCH (TORCH) institutes arranged by the foundation and taught by teams of master teachers selected from the four-week summer institute participants. The one-week institutes on a particular topic begin the summer after that topic is the focus at a summer leadership institute, and they continue to be scheduled in subsequent summers. For example, through the summer of 1992, more than forty one-week statistics institutes have been held since the summer Leadership Institute on Statistics in 1984. During the summer of 1992, there were fifty-four one-week mathematics institutes on various topics at sites around the United States. University professors or district mathematics coordinators make arrangements for these one-week institutes and often reap unexpected rewards.

> "It was impressive for me to see superb teachers in action and to watch our local teachers absorb it all. You (the master teachers) clearly helped make a very good group of teachers better teachers. Much of your approach and style I, too, can use. In short, you have made me a better teacher as well."

The one-week institutes administered by the Woodrow Wilson Foundation are funded in part by the sites, often using corporate, foundation, and Dwight D. Eisenhower funds.

EVALUATION OF THE PROGRAMS

The power of the Woodrow Wilson program lies in the extent of its influence. Fifty teachers attend the four-week summer institute. These individuals work with other teachers throughout the country through outreach projects and one-week institutes. Even allowing for teachers who attend more than one institute and those who retire, the number of students affected by the program was well over 2.5 million in 1991.

Each of the quotations appearing in this article is taken from a narrative evaluation of a four-week summer institute, an outreach project, or a one-week institute. Such evaluations, a regular part of any Woodrow Wilson activity, provide the impetus for continuing this model for change.

For several years, an independent evaluator has analyzed the accumulated data on the effectiveness of the one-week institutes. More than 85 percent of the participants reported that the approach of teachers teaching teachers was of more value than other in-service experiences. High ratings were given to the materials demonstrated and to the teaching techniques modeled.

Results of the National Leadership Program for Teachers—for the master teachers as well as for the recipients of their outreach efforts—include an expanded vision of what teaching can be, a renewal of teachers' enthusiasm for their profession, the development of a strong network of supportive colleagues, and an impetus for continuing to learn. The Woodrow Wilson Foundation, through its National Leadership Program for Teachers, is one of the change agents giving life to the vision of the NCTM *Standards* documents.

> "Within only a short time the first morning, I discovered that I was at something much greater than I had envisioned. By the end of the experience, I had a totally new outlook on what I can do. It is as if I have crossed over the bridge to the future. I know that I am only over the bridge and that it will take time and effort to go forward, but at least I am headed in a better direction. Thank you."

REFERENCES

National Council of Teachers of Mathematics. *Professional Standards for Teaching Mathematics*. Reston, Va.: The Council, 1991.

Steen, Lynn Arthur, ed. *On the Shoulders of Giants: New Approaches to Numeracy*. Washington, D.C.: National Academy Press, 1990.

22

Teachers Empowering Teachers: A Professional-Enhancement Model

Martha Wallace
Judith Cederberg
Richard Allen

B OTH the *Curriculum and Evaluation Standards for School Mathematics* (National Council of Teachers of Mathematics [NCTM] 1989) and the *Professional Standards for Teaching Mathematics* (NCTM 1991) view the learner as an active constructor of knowledge who integrates new concepts into an existing knowledge structure. Recent research in cognitive psychology and mathematics education has led to a vision of teaching as a process in which teachers find ways to help learners build their own knowledge. Teachers can no longer merely supply correct information to students who are expected to absorb it passively; they must act as catalysts and facilitators of learning (Linn 1986).

For teachers to engage their students' intellects and interests to induce the active learning advocated by the *Curriculum and Evaluation Standards,* they must learn the new ways of teaching promoted by the *Professional Teaching Standards*: (1) selecting appropriate tasks and providing opportunities to deepen students' understanding, (2) orchestrating classroom discourse, (3) using technology to pursue mathematical investigations, (4) finding—and helping students find—connections to other knowledge, and (5) guiding individual and group work (NCTM 1991).

Saint Olaf College has been helping teachers learn these new ways of teaching through a project, Teachers Empowering Teachers: Computers in Geometry

The project described in this article was supported by the National Science Foundation through Teacher Enhancement Project no. TPE8955118. The opinions expressed are those of the authors and not necessarily those of the foundation.

The authors wish to thank Dan Chazan, of Michigan State University, who acted as a consultant on the project. This article would not have been possible without the fifty secondary school geometry teachers who participated in the project and shared their wisdom with us and each other.

Earlier versions of portions of this paper have been presented at the Fourth Annual International Conference on Technology in Collegiate Mathematics (Portland, Oreg., November 1991) and at the Seventh International Congress on Mathematics Education (Quebec City, Canada, August 1992.)

Classrooms, that envisions teachers as constructive learners in the same way in which the *Curriculum and Evaluation Standards* views students as learners. That is, the project assumes that teachers will learn new ways of teaching mathematics only if their intellects are engaged and only if they have opportunities to construct their own knowledge and to integrate that new knowledge into their existing knowledge structures. The central aim of the project, based in the mathematics department of the college, is to help secondary school geometry teachers use computer technology to make the teaching and learning of geometry an engaging, dynamic activity consistent with the recommendations of the *Curriculum and Evaluation Standards*.

GOALS AND ASSUMPTIONS

The project has two goals. The first is to help geometry teachers develop the knowledge, skills, and confidence necessary to use computer-based tools to transform their geometry classrooms into mathematical communities where students explore, conjecture, verify, and communicate mathematics. In this community, the teacher is a partner in inquiry instead of the sole authority for correct answers. The second goal is to enable these teachers to impart their expertise to other teachers and to student-teachers.

The project is guided by five basic assumptions: (1) computer-based technology can stimulate significant improvements in the way in which geometry is taught and learned; (2) teachers as well as students must construct their own knowledge through active involvement; (3) the content of effective in-service programs must include subject-specific as well as pedagogical knowledge; (4) practicing teachers must play a major role in the planning and implementation of any effort to improve classroom teaching; and (5) any project that leads to the improvement of classroom teaching must include extended support for teachers. On the basis of these assumptions, the project assists teachers in developing their own strategies for using computer technology to enhance their geometry teaching. The primary vehicle used in this process is the teaching scenario, which is a set of integrated lesson plans incorporating computer technology wherever appropriate. All the teacher-participants in the project have written their own scenarios and are teaching and critiquing both their own and other teachers' scenarios.

PROJECT DESIGN

The project makes use of an "expanding network" model. Beginning with three college instructors, the network expands to ten master high school geometry teachers, then to forty less-experienced teachers, and finally to colleagues in each of the fifty teachers' schools. As the description below indicates, experts in each level of the network serve those in the next level, first as coaches and catalysts for learning, then as mentors, and finally as colleagues. The activities described occurred as indicated in the time line in figure 22.1.

Level 1: College Faculty

The first level of the network consists of the three college instructors, who are codirectors of the National Science Foundation grant that supports the project. Although the college instructors are all members of the mathematics department at Saint Olaf, each has a different specialty. The instructors contribute expertise in the areas of mathematics education, geometry, and computer science. They serve as teachers, coaches, facilitators, and mentors for the secondary school teachers who form the expanding network.

Spring 1990	Selection of TAGs*, planning
Summer 1990	TAG Workshop 1—two weeks, eight days of scenario writing
	TAG Workshop 2—two days
Fall 1990	TAGs taught two scenarios, college staff observed
January 1991	TAG Workshop 3—two days
Spring 1991	TAGs taught fractal scenario, college staff observed, selection of ENTs**
Summer 1991	ENT Workshop, TAGs assisting—two weeks
Fall 1991	ENTs taught two TAG scenarios, TAG mentors observed
January 1992	ENT and TAG Workshop—two days
Spring 1992	ENTs teaching ENT scenarios, preliminary student evaluation
Summer 1992	National Conference and Scenario Workshop

* TAG = Teacher Action Group
** ENT = Extended Network Teachers

Fig. 22.1. Time line

Level 2: Teacher Action Group

As soon as the initial plans were formulated, a group of ten secondary school geometry teachers, called the "Teacher Action Group," was included in the planning. These ten teachers were invited to participate because of their acknowledged interest, expertise, professional involvement in geometry teaching, proximity to Saint Olaf, and the support they received from their principals. During the first year, these teachers acted primarily as learners while attending a two-week workshop taught by the college instructors. They developed scenarios, and taught their scenarios under the mentorship of the college instructors.

As students during the first summer workshop, the ten teachers in the Teacher Action Group explored aspects of modern geometry, including finite and non-Euclidean geometries and analytical approaches to transformational geometry. The teachers studied the NCTM Standards and other recommendations for reform and considered issues in, and guidelines for, integrating computer technology into geometry teaching. The teachers also investigated the capabilities and limitations of several geometry exploration software packages (Geometric Supposers and Transformations, Geometer's Sketchpad, LogoMath, René's Place and Euclid's Toolbox, and Geometry One and Two). Each teacher targeted a topic in the fall geometry curriculum, chose a software package that would fit the teaching goals and the available hardware, and planned a scenario designed to engage students in exploring the chosen geometry topic. (The majority of the teachers chose the Geometric Supposers, often because their schools had only Apple II computers available for the students' use. Two selected the Geometer's Sketchpad for primary use, and others used it for demonstration.) Following the two-week workshop, the teachers spent an additional eight days developing their scenarios, which were then critiqued by the group during a two-day workshop in late summer.

As the ten teachers in the Teacher Action Group taught their own and others' scenarios, they evaluated the scenarios and monitored their own thinking and learning in the process of teaching familiar content in new ways. They shared insights at a two-day meeting in January and, guided by their own experiences, assisted in the planning for the next phase of the project. During the spring, these teachers taught a scenario on fractal geometry that had been written by the college faculty. As they prepared for and taught the fractal scenario, the teachers reflected on the preparation they needed to teach unfamiliar mathematical content.

Throughout the first year of the project, the ten teachers continually analyzed their experiences and reactions in order to gain insights that would prepare them to be teachers and mentors to the next group of teachers. After discussion, the teachers made recommendations to the college faculty about selection criteria, the workshops' content and process, and necessary follow-up activities.

Level 3: Extended Network Teachers

While the first ten teachers were teaching and evaluating their scenarios during the fall, brochures were sent to schools throughout the state to solicit applications for the next level of the network, the "Extended Network Teachers." Applicants detailed their previous professional involvement, defined their personal goals for participation in the project, and provided evidence of administrative support for their project activities. The professional qualifications of the successful applicants covered a wide range. Some had backgrounds that could have qualified them as members of the Teacher Action Group, and others were first-year teachers. The qualifying traits included at least a beginning understanding of the vision of the NCTM's *Curriculum and Evaluation Standards,* a view of computer use as a means to engage students actively in learning at a level beyond rote skills, and a willingness to try new ways of teaching. Forty teachers from throughout the state were selected to be members of the Extended Network.

These forty Extended Network Teachers began their activities with a two-week workshop during the second summer of the project. At the suggestion of members of the Teacher Action Group, the forty new members of the project were introduced to the philosophies of the project through a preworkshop reading assignment (Chazan and Houde 1989). The workshop itself was more structured than the initial Teacher Action Group workshop for two reasons. First, because the Extended Network Teachers would not have the extra eight days after the workshop to work on developing their scenarios or the late summer meeting to share and critique, it was deemed essential that each complete a scenario in strong first-draft form before leaving the workshop. Second, it did not seem wise to involve a more mixed group of teachers in the extended open-ended exploration used in the initial Teacher Action Workshop during which an attempt was made to have the first ten teachers discover as much knowledge as possible for themselves. Thus, the Extended Network Teachers were given guidelines for developing scenarios (see fig. 22.2) as well as copies of the scenarios developed by the original ten teachers. Furthermore, the forty new teachers were exposed only to those pieces of software deemed most appropriate by the first ten teachers, primarily the Supposers and Transformations (Sunburst) and Geometer's Sketchpad (Key Curriculum Press).

The ten members of the Teacher Action Group played a major role in this workshop by introducing both the selected software and their own scenarios. They also served as classroom assistants in the modern geometry class, as discussion leaders in the mathematics education sessions, and as staff in the scenario-writing laboratories, helping their forty new colleagues in tasks ranging from goal definition to word processing the final product. Because most of the fifty teachers in the project stayed in a college dormitory, much informal discussion and guidance took place at night. By the end of the workshop, the fifty teacher-participants had formed fast bonds, and each of the original ten teachers was assigned as a mentor to work closely with four members of the Extended Network.

During the school year following the summer workshop for the Extended Network Teachers, each of these forty teachers used four scenarios—two scenarios developed by members of the Teacher Action Group, their own scenario, and the fractal scenario. As they prepared for and taught each of the scenarios, the teachers evaluated the process and the materials, noting how they adapted scenarios written by others and how they revised their

What is a scenario?	An integrated set of lesson plans (a dynamic unit plan) designed to elicit active student learning of a coherent topic in geometry. Your scenario is your plan for teaching a topic in geometry in a manner that will empower your students mathematically.
What should it cover?	It may cover roughly the same topic as a chapter in your textbook; it may combine traditional geometry content in a way that is different from the textbook organization; or it may cover an enrichment topic not taught in the traditional course.
How long should it be?	In most instances, the scenario will take two or three weeks to teach.
What are the goals?	It should be consistent with the vision of the NCTM *Curriculum and Evaluations Standards for School Mathematics* and with the conception of the learner as an active participant in the learning process. It should encourage higher-order thinking, conceptual understanding, and student-constructed knowledge.
What is the role of computers?	Computer tools should play a significant role in the scenario, but not every lesson needs to be computer based. Other active-learning tools, such as constructions, paper folding, large- and small-group discussions, reading, and writing, should also be incorporated where appropriate.
What should it contain?	1. An introduction, including the topic, and time frame; how and when the topic fits into the curriculum; and the goals, prerequisites, and resources (books, software, hardware, other tools, etc.) 2. Daily lessons, including preparation, resources, activities, homework, and culminating activities where appropriate 3. An assessment or evaluation instrument

Fig. 22.2. Scenario-development guidelines

own scenarios. Each member of the Extended Network kept in contact with a mentor (assigned according to geographic proximity, compatibility, and common software interests), who provided feedback, teaching ideas, and software information. The entire network continued to be maintained by periodic newsletters, fall and spring meetings at state mathematics conferences, and a two-day workshop in January.

Level 4: The Growing Network

The final level of the network is less formal than the first three. Several of the project's fifty teachers are guiding other teachers in their schools who are also using the project scenarios. Some are supervising student-teachers as they teach scenario-based units. Members of both the Teacher Action Group and the Extended Network have made several presentations on project activities at state mathematics meetings. The college has made computer accounts available to the project participants, primarily so that they can communicate with each other and access national geometry networks by electronic mail.

A Scenario Workshop was held the third summer of the project for geometry teachers and mathematics educators from throughout the country. During this workshop many of the teachers demonstrated geometry-exploration software, conducted simulated classroom sessions using lessons from their scenarios, and led discussions on their experiences in integrating computer tools into their geometry teaching. Most of the project teachers brought colleagues with them to the Scenario Workshop.

Immediately preceding the Scenario Workshop, a national conference examined issues and showcased projects in which computer tools were used to revitalize secondary school geometry. Prominently featured in the conference were demonstrations and laboratory sessions highlighting the software used in the project as well as emerging and newly available geometry-exploration software, such as Cabri, Geometry Grapher, GeoExplorer, Geometry Inventor, and superSupposer. The conference helped to build a mental framework for the teachers who participated in the demonstrations and discussions that followed in the Scenario Workshop. The project participants and their school colleagues had been especially encouraged to attend the national conference as well as the Scenario Workshop.

SCENARIOS

This project is built on a process-oriented model in which teachers work together to construct and augment their knowledge of ways to use dynamic teaching methods to engage students in exploring geometric concepts and generalizations. The scenarios became the concrete vehicles by which the teachers explored, reevaluated, and revised the content, their teaching methods, and their beliefs about geometry and geometry teaching. Indeed, clarifying the concept of a scenario and organizing and writing scenarios became the first major tasks of each group of teachers.

The curricular emphasis in most of the scenarios was quite traditional for several reasons: (1) Because most teachers entered the project with very little experience in using computer microworlds and had only recently begun to think about the kind of active, student-centered instruction advocated by the NCTM's *Curriculum and Evaluation Standards,* it seemed important to start improving traditional, familiar topics. As one of the members of the Teacher Action Group said, "We need to gain experience using these tools within the current curriculum in order to help form the future curriculum." (2) Teachers were constrained by state, district, and school guidelines for their geometry curricula. (3) Several of the project schools were beginning to use Chicago Project Geometry textbooks (Coxford, Usiskin, and Hirschhorn 1991) concurrently with project activities, and teachers were reluctant to modify the Chicago-recommended scope or sequence. (4) Some school colleagues and administrators were suspicious of radical change.

For these reasons, many project teachers chose to move carefully by starting with traditional curriculum topics. Their scenario topics included angles, area, circles, constructions, lines, polygons, quadrilaterals, similarity, transformations, triangles, visualization, and an eighth-grade geometry unit. However, all these scenarios incorporated the use of computer-assisted tools and emphasized students' conjecturing. A list of the scenario titles, along with the associated hardware and software, is available from the authors.

NEW BELIEFS, NEW QUESTIONS

Although the teachers in this project expressed understanding of, and agreement with, the premises of the project, it took some time for them to alter their beliefs about teaching geometry. The early drafts of the first scenarios often treated computer tools more as instructional ends than as means to help students construct their own knowledge. However, after one semester of teaching the scenarios, the teachers began to view the use of computerized tools within a larger framework. During the first January meeting, the Teacher Action Group assessed their own experiences in preparation for planning the summer workshop for the Extended Network Teachers. They agreed on the following basic premises:

- Computer technology is only one of the tools that can be used to promote active learning and thinking. Other tools for active learning include construction tools, calculators, measuring devices, manipulatives, and paper folding.
- Teachers should use whichever tools work best for each particular learning goal and should equip students to choose and use whichever tools work best for each particular mathematical goal.
- Appropriate computer use in geometry classrooms requires and generates a new classroom ecology—cooperative learning, laboratory methods, group discussions, and so on.
- Appropriate computer use in geometry classrooms requires and encourages higher-order thinking skills such as conjecturing, problem posing, problem solving, verifying, and visualizing.
- Proof is only one part of geometry.

As the project teachers became more comfortable with new ways of teaching, they uncovered questions that had not occurred to them before. To give some flavor of their concerns, a sample of the questions they generated during the first January meeting is listed in figure 22.3.

Conjecturing

What is it?

Is it our only goal?

When during the year should it be introduced?

How do teachers deal with a polarized class in which some catch on quickly (especially to developing conjectures) and some never do?

What do teachers do if very few students get a conjecture or get only "trivial" conjectures?

How can teachers get students to feel ownership for conjectures developed by others?

Laboratories

What do teachers do if they are limited to using only demonstrations in teaching?

How can teachers best combine demonstration and laboratory activities?

How do teachers handle absences on laboratory days?

How do teachers handle inadequate conditions?

How do teachers solve problems of vandalism?

How do teachers handle transporting classes of students between a classroom and a laboratory?

How do teachers manage time?

How do teachers carry on noncomputer instruction in a laboratory setting?

How does the classroom climate change?

How do the teacher's and students' roles change?

What if the lesson doesn't last long enough? Lasts too long?

How do teachers get students to record data?

Curriculum

How do teachers integrate new ideas with the textbook sequence (especially if students are developing conjectures that are in much later chapters)?

Since inserting new material requires extra time (especially during the first year), what gets lost? What must be saved?

How do teachers balance enrichment exercises with guiding discovery?

How much computer busywork is done versus "real mathematics"?

Assessment

How do teachers hold students accountable?

Are students evaluated and how?

What kind of testing (if any) should be done on computer use and how?

Are these efforts worthwhile? How do we know?

Fig. 22.3. Teacher-generated questions

Enlightened by the first year of the project, we were able to plan the first workshop for the Extended Network Teachers to anticipate and address such questions. As a result, most of the members of the Extended Network were able to focus on facilitating active learning more quickly than the members of the Teacher Action Group. However, several of the questions remained. When the members of the Extended Network were asked to indicate issues that needed discussion at the second January workshop, they listed such questions as these:

- What can be eliminated from the curriculum to find time for inquiry teaching?
- How does one plan opportunities for inquiry learning? How do I get students to make nontrivial conjectures?
- How do I assess inquiry learning?
- How can I use a one-computer classroom effectively?

RESULTS

One test of a professional development program is whether it effects long-lasting change in teachers' beliefs and behaviors. A preliminary evaluation of the Teachers Empowering Teachers project shows that the participating teachers are continuing to explore new ways of teaching geometry. During the second year of the project, the Teacher Action Group continued to use and revise the scenarios that they and their colleagues had developed during the first year. Several adapted for their own use some of the scenarios developed by the Extended Network Teachers. Several teachers from both the first and the second groups have also written new scenarios.

One of the results of the project was the nurturing and increasing of the teachers' awareness of their importance and their potential impact as professional change agents. Since the first group of teachers was selected partly because of their professional activity, it was not surprising that they shared the knowledge and experience gained through the project with other teachers by speaking at state professional meetings and conducting workshops. However, the Extended Network Teachers also expanded their professional networks by giving talks and workshops—some for the first time. A supervisor of one of the Extended Network Teachers noted that the teacher had always been known as an accomplished classroom instructor but for the first time was reaching out to other teachers. This teacher has also been conducting several in-service workshops in the district.

Other aspects of the project's effects still need to be evaluated. In particular, we have only begun to evaluate the ways in which students' learning in geometry is enhanced by the integration of computer-assisted tools. This is an important question, and extensive work on this issue should constitute a new field of research.

CONCLUSION

The *Curriculum and Evaluation Standards* (1989) calls for the increased use of computer technology in all areas of mathematics teaching and for teaching in such a way that students will gain mathematical power. The model used in the Saint Olaf project offers ideas for developing teachers' power. Such power helps teachers (1) acquire the knowledge and self-confidence necessary to use technology appropriately to enable students' active learning and (2) develop the ability to impart that knowledge and self-confidence to other teachers. We have found that teachers can learn to integrate computer technology into their geometry teaching. They can develop new styles of teaching to elicit higher levels of student thinking, and they can find ways to encourage and accommodate new classroom structures and ecologies. These changes in teachers' behavior result from changes in their knowledge and beliefs. Achieving and sustaining these results requires extensive opportunities for teachers to explore and construct their own knowledge with the encouragement of extended support networks.

REFERENCES

Chazan, Daniel, and Richard Houde. *How to Use Conjecturing and Microcomputers to Teach Geometry*. Reston, Va.: National Council of Teachers of Mathematics, 1989.

Coxford, Arthur F., Zalman Usiskin, and Daniel Hirschhorn. *The University of Chicago School Mathematics Project Geometry*. Glenview, Ill.: Scott, Foresman & Co., 1991.

Linn, Marcia. *Establishing a Research Base for Science Education: Challenges, Trends, and Recommendations*. Washington, D.C.: National Science Foundation, 1986.

National Council of Teachers of Mathematics. *Curriculum and Evaluation Standards for School Mathematics*. Reston, Va.: The Council, 1989.

———. *Professional Standards for Teaching Mathematics*. Reston, Va.: The Council, 1991.

GEOMETRY SOFTWARE

Cabri. Pacific Grove, Calif.: Brooks/Cole Publishing Co., 1992.

Euclid's Toolbox. Menlo Park, Calif.: Addison-Wesley Publishing Co., 1992.

GeoExplorer. Glenview, Ill.: Scott, Foresman & Co., 1992.

Geometer's Sketchpad. Berkeley, Calif.: Key Curriculum Press, 1991.

Geometric Connections: Transformations. Pleasantville, N.Y.: Sunburst/Wings for Learning, 1990.

Geometric Connections: Coordinates. Pleasantville, N.Y.: Sunburst/Wings for Learning, 1991.

The Geometric preSupposer. Pleasantville, N.Y.: Sunburst/Wings for Learning, 1991.

The Geometric superSupposer. Pleasantville, N.Y.: Sunburst/Wings for Learning, 1992.

The Geometric Supposer Series (Circles, Quadrilaterals, Triangles). Pleasantville, N.Y.: Sunburst/Wings for Learning, 1986–1989.

Geometry Grapher. Boston: Houghton Mifflin Co., 1990.

Geometry Inventor. Pleasantville, N.Y.: Sunburst/Wings for Learning, 1992.

Geometry One: Foundations. Armonk, N.Y.: IBM (Wicat Systems, Inc.), 1990.

Geometry Two: Proofs and Extensions. Armonk, N.Y.: IBM (Wicat Systems, Inc.), 1990.

LogoMath: Tools and Games. Portland, Maine: Terrapin Software, 1990.

René's Place. Menlo Park, Calif.: Addison-Wesley Publishing Co., 1992.

23

The Kentucky K–4 Mathematics Specialist Program

William S. Bush

IN 1989, the United States began a national reform in mathematics education initiated largely by the *Curriculum and Evaluation Standards for School Mathematics* from the National Council of Teachers of Mathematics (NCTM) (1989) and *Everybody Counts* from the Mathematical Sciences Education Board of the National Research Council (1989). These documents called for sweeping changes in both mathematics curricula and teaching approaches. Undoubtedly, many if not most teachers of mathematics across the country realized that they needed to learn new mathematics content as well as new ways to teach mathematics. Appropriate, sustained support for teachers of mathematics through professional development was necessary if these changes were to be realized within the ten-year timetable suggested by the NCTM *Curriculum and Evaluation Standards*. Clearly, to meet these new professional development demands would also require new kinds of collaboration among many groups of people.

In Kentucky, professional development for mathematics teachers encountered several obstacles. First, professional development at the district level was restricted by limited physical and human resources. In 1989, one-third of the 1900 elementary school teachers surveyed had no access to manipulative materials for teaching mathematics (Bush, Vice, and Gilbert 1989). The few experts in mathematics education in the state were inaccessible to the 176 school districts in Kentucky, especially the small rural districts in eastern and western Kentucky. All but two school administrators in Kentucky were generalists and could provide little professional development in mathematics education.

In 1989, the mathematics education community sought a solution to the problem of furnishing quality professional development to Kentucky's teachers of mathematics. The community decided that the first priority in implementing the

The program reported in this manuscript was supported by the National Science Foundation under grant no. TPE-9147162. All opinions expressed are solely those of the author.

Standards was support for teachers of young children. Therefore, the mathematics education leaders developed a statewide program to prepare K–4 classroom teachers across Kentucky as mathematics specialists. A steering committee of eight university mathematics educators and mathematicians, two classroom teachers, one school administrator, and a staff member of the Kentucky Department of Education subsequently received a grant from the National Science Foundation to implement a large-scale professional development program.

PROGRAM GOALS, SCHEDULE, AND STRUCTURE

The Kentucky K–4 Mathematics Specialist Program began in May 1990 with the following goals:

- To establish a comprehensive statewide network of K–4 mathematics specialists to serve as resource persons in approximately one out of every five elementary schools in Kentucky
- To align mathematics content and teaching practices in grades K–4 in Kentucky with the recommendations in NCTM's *Curriculum and Evaluation Standards* and the Kentucky Education Reform Act of 1990
- To provide opportunities for communication and collaboration among classroom teachers, school supervisors, teacher educators, and college mathematicians regarding mathematics content and teaching

The steering committee divided Kentucky into eight regions; each region included one state university. The regions were established so that between 100 and 150 public and private elementary schools were included in each. The demographics of the regions varied greatly. The region associated with the University of Louisville was primarily urban and suburban and contained teachers from a large public school system (Jefferson County Schools), a large Catholic diocese, and several small private schools. The region associated with Northern Kentucky University served small public and private districts in the suburbs of Cincinnati, Ohio. The region associated with the University of Kentucky included urban and suburban schools in Lexington, the second largest city in the state, as well as many smaller rural districts in south-central Kentucky. The remaining five regions contained primarily small rural districts in the Appalachia region of eastern Kentucky and the rolling farmland of western Kentucky. One training site was selected in each region so that teachers were located within a two-hour drive to the site. The University of Kentucky region selected two sites, with one serving as a pilot site.

The steering committee selected teams of six persons, including a mathematics educator, a mathematician, three exceptional K–4 teachers, and an instructional supervisor, to represent each region. The K–4 teachers applying to the program were required to complete an extensive application form. Those classroom teachers selected received over $3500 in stipends, two microcomputers, mathematics software for grades K–4, a $500 manipulatives kit, and a classroom set of calculators.

In summer 1990, all forty-eight team members from the eight regions attended a three-week institute at the University of Kentucky. At the beginning of the institute, the team members received the most up-to-date information on K–4 mathematics and teaching K–4 mathematics from an array of national experts. In the last week of the institute, the team members were divided into four content groups: geometry, number sense, measurement, and statistics. These groups developed drafts of modules for each content area. Each module included a rationale, a theoretical model, and exemplary activities for teaching the content. The groups were also required to incorporate problem solving, manipulatives, small-group work, and technology into the activities in their modules.

During the 1990–91 school year, the twenty-four classroom teachers on the regional teams field-tested the ideas and activities developed at the institute with their K–4 students. The university faculty field-tested the module ideas and activities in their mathematics and methods courses. The regional teams met monthly to share ideas and discuss the activities.

In fall 1990, writing teams organized by the director of the program created modules on developing lessons, using alternative assessment, involving parents, and conducting workshops. In spring 1991, the University of Kentucky regional team conducted a pilot version of the mathematics specialist seminars for thirty K–4 teachers in the central Kentucky area. These seminars were videotaped, and other regional team members attended them regularly.

Also in spring 1991, applications for mathematics specialists were sent to every elementary school in the state. Commitment forms asking districts to pay a stipend of $300 and to purchase a five-hundred-dollar manipulatives kit for each teacher chosen were sent to the local Eisenhower Program coordinators across the state. The teachers who applied and received support from their administration were screened by the steering committee. Of the 356 teachers who applied, 240 were chosen to attend the seminars.

Forty-nine regional team members (a teacher was added to one regional team) returned to the University of Kentucky for a two-week institute in summer 1991. During this time, the content groups revised the modules on the basis of information gathered during field-testing and of results from the spring pilot seminars at the University of Kentucky. The revised modules (Bush, Nichols, and Prasse 1991) were edited, published, and distributed to all the regional teams in August.

THE MATHEMATICS SPECIALIST SEMINARS

Beginning in fall 1991, each regional team, using the revised modules, trained the 240 selected K–4 teachers in fifteen three-hour seminars. The seminars were conducted one night a week from September through December at each regional site. The participating K–4 teachers had the option of receiving graduate credit for the seminars if they were willing to do extra work.

The first module, Opening Night, presented an overview of the program, the NCTM Standards, and the Kentucky Education Reform Act of 1990. Seminars 2 through 9 supplied rationales, theoretical models, and activities for teaching the four content areas in grades K–4. The theoretical model for geometry was based on the van Hiele developmental levels for learning geometry. The number-sense module included number concepts, place-value concepts, counting strategies, number composition and decomposition, basic facts, estimation, and mental computation. The theoretical model for number sense furnished direction in using concrete materials and in moving from concrete materials to abstract concepts. The measurement module offered five stages for teaching measurement: (1) identifying attributes, (2) making comparisons, (3) using nonstandard units, (4) using standard units, and (5) using formulas. The statistics module furnished a developmental model for assisting young students with data collection, organization, representation, and interpretation.

The focus of the seminars changed on Nights 10, 11, and 12. During these seminars, the teachers discussed how to construct and plan for meaningful mathematics lessons and how to integrate problem solving, manipulatives, small groups, calculators, and computers into their lessons. The regional teams encouraged the teachers to reflect on the activities and theoretical models presented during Nights 2–9. On Night 13, the teachers discussed the Kentucky assessment plan (Kentucky Department of Education 1991) and how to develop and use such alternative forms of assessment as performance-based tasks, portfolios, writing, interviews, observations, and projects. On Night 14, the teachers brought school administrators (principals, instructional supervisors, superintendents), school board members, and parents to the seminars. The guests were given an overview of the program and NCTM's Standards. The parents and administrators worked in groups with their associated teachers to develop action plans for getting parents involved and for using the newly trained mathematics specialists in their respective districts. On Night 15, the teachers developed guidelines for preparing and conducting workshops for colleagues.

Small-group work was the predominant format during the seminars. Lecture was used rarely and only when information outside the teachers' experience (e.g., information about the program, the NCTM Standards, or the van Hiele levels) was presented. During Nights 2–9, the mathematics specialists were engaged in group mathematics activities followed by reflective discussion about the activities. During Nights 10–15, the mathematics specialists were asked questions regarding classroom practices and beliefs. They worked in groups to reach a consensus and formulate answers. Although the team members made sure that certain issues or concerns were addressed, most of the information and discussion came from the participating teachers themselves.

The modules were constructed so that the regional teams could adapt the content to particular participants' and regional needs. For example, in the urban regions where participants had attended a significant number of workshops and professional development programs, the team conducted fewer activities but allowed more time for sharing, discussion, and reflection. In rural regions, where some participants traveled two hours to the seminars, the team members incorporated more hands-on activities throughout the evening to maintain a high level of interest.

The mathematics specialists not taking the seminars for graduate credit were asked to spend three hours each week on assignments outside the seminars. The assignments included a weekly journal entry addressing a specific question or issue, a classroom assignment requiring them to field-test an activity with their students, a biweekly set of challenging mathematics problems related to the content being studied, and a specific reading assignment. The mathematics specialists taking the seminars for graduate credit were asked to spend six hours each week on assignments outside the seminars. In addition to the assignments above, they were required to write about their classroom assignments and to spend more time on the set of challenging mathematics problems.

FOLLOW-UP

Each of the forty-nine regional team members were assigned 5 to 7 of the 270 mathematics specialists (30 trained in spring 1991 and 240 trained in fall 1991). The team members met with their assigned mathematics specialists once a month during spring and fall 1992. At these meetings, teachers shared ideas, discussed problems, and planned district workshops. The team members, in essence, served as peer coaches for the mathematics specialists and were available for assistance one year after the seminars.

During summer 1992, the forty-nine team members again attended an institute at the University of Kentucky. At this one-week institute, they worked with consultants from the Used Numbers Project to learn more about teaching data analysis in grades K–4. The team members, in turn, developed (1) performance-based tasks involving data analysis for children in grades K–4 and (2) data analysis lessons that can be incorporated into thematic units about the environment, movement, communities, and children around the world. During fall 1992, the team members distributed these tasks and lessons to their assigned mathematics specialists through workshops and monthly meetings. The mathematics specialists, in turn, shared the materials with other teachers in their districts through local workshops.

Additional funding from the Exxon Education Foundation and the Kentucky Department of Education permitted another round of mathematics specialist seminars to be conducted by regional teams at six sites across the state in fall 1992. An additional 120 K–4 teachers in Kentucky attended these seminars to become mathematics specialists.

PROGRAM EVALUATION

The program also had an extensive research-and-evaluation component. Two research teams of university faculty were formed to determine the effects of the program on teachers and children across Kentucky. The first team, called the "microresearch team," visited the classrooms of the original twenty-four teachers selected as regional team members, as well as a random sample of twenty teachers chosen as mathematics specialists. During the classroom visits, the researchers observed a mathematics lesson, interviewed the teacher, and interviewed five or six children. The teachers in the research component were also required to complete a checklist of their daily activities (e.g., use of manipulatives, problem solving, small-group work, calculators, computers). They also responded to essay and journal questions posed throughout the year.

The second team, called the "macroresearch team," developed a survey to assess teachers' awareness of educational reform, knowledge of mathematics, beliefs about mathematics, and teaching practices. This survey was administered to the 240 mathematics specialists at the beginning and end of their fall 1991 seminars. These teachers were asked to complete the survey once a year during the subsequent two years of the project. The results of the survey gave some indication of the effects of both the seminars and the program.

THE PROGRAM IMPACT

The success of the Kentucky K–4 Mathematics Specialist Program has far exceeded the goals and expectations of the steering committee.

Establishing a Network

The first goal, establishing a network of mathematics specialists, was clearly met. The twenty-five teachers on the regional teams have become very active in professional mathematics education activities. All conducted at least one local or state mathematics workshop. Eight presented at state or regional professional meetings. Thirteen were appointed to Kentucky Department of Education committees. Three left their classrooms to become mathematics resource teachers in their school districts, and three have become principals.

The 390 teachers prepared as mathematics specialists and the 25 teachers on the regional teams represented 30 private schools and 141, or 80 percent, of Kentucky's 176 public school districts. These 141 districts employ approximately 90 percent of the state's public elementary school teachers. Most of the 270 mathematics specialists who attended seminars in 1991 conducted workshops in their districts during spring and summer 1992.

Prior to the program, the mathematics education community of Kentucky was composed primarily of university faculty, secondary school teachers, and middle grades teachers. Few elementary school teachers in the state were actively involved in statewide mathematics education activities. As a direct result of the K–4 program, elementary school teachers have a substantial representation in the mathematics education community as well as in statewide activities in mathematics education. For instance, committees formed by the Kentucky Department of Education to develop statewide mathematics goals, assessment items, and curricula included many of the program's regional team members and mathematics specialists. Also, team members and mathematics specialists have made presentations at regional and state mathematics education conferences.

Aligning with the Standards and the Kentucky Education Reform Act

The second goal, aligning curriculum and teaching with the NCTM's Standards and the Kentucky Education Reform Act of 1990, was also accomplished. The four content and teaching-process areas of the program were based on recommendations from the *Curriculum and Evaluation Standards*. Following the national reform movement, the Kentucky Department of Education developed a curriculum framework (Kentucky Department of Education 1990, 1992) and valued outcomes for students in grades 4, 8, and 12. The mathematics components of the outcomes were based on NCTM's Standards. School districts develop local curricula based on the framework. Therefore, with mathematics specialists available to develop district mathematics curricula, practically all district mathematics curricula in the state will eventually align with the Standards.

The program also assisted in implementing two mandates of the Kentucky Education Reform Act of 1990: alternative assessment and nongraded primary school education. The seminars included performance-based tasks and portfolios, specific components of the Kentucky Instructional Results Information System (KIRIS) (Kentucky Department of Education 1991). In a nongraded primary school environment, K–4 teachers are required to make decisions about placing students in groups and providing developmentally appropriate activities for them. The theoretical models of the seminars were designed to help teachers make such decisions. The teachers who completed the program and understood the theoretical models reported having little difficulty in teaching mathematics in the nongraded environment.

Establishing Collaboration

The third goal, establishing collaboration among groups, was very successful. The training teams of university faculty, classroom teachers, and school administrators worked exceptionally well together. The university faculty assured that the modules were mathematically and pedagogically sound, and the classroom teachers assured that the modules presented models and activities appropriate for K–4 teachers and students. The regional training teams also worked well together. The mathematics educators typically posed questions and focused on pedagogical issues; the mathematicians tended to pose questions about the mathematics content of the activities; and the classroom teachers often shared stories and products from their own classroom experiences. The seminars were deemed beneficial because they were produced by professionals with varying strengths and perspectives.

The use of regional teams to conduct seminars enhanced the quality of the program. Local team members understood the constraints, problems, attitudes, environment, and culture of the teachers in their region and could address them appropriately. The team members also understood and used the language of their regional colleagues. Simply, put, participating teachers were more receptive to advice and suggestions from others in their situation.

The substantial statewide impact of the program was due to the collaboration and support of the 141 school districts. Eighty percent of the school districts supported at least one teacher for the program at a cost of $800 for each teacher ($500 for a manipulatives kit and $300 for a stipend). Local Eisenhower funds were often used for these contributions. In contrast to the results of the 1989 survey by Bush, Vice, and Gilbert (1989), at least 415 K–4 teachers now have manipulatives kits in their classrooms. At least 200 additional kits were purchased by districts within the first year after the fall 1991 seminars.

Several corporations, including Apple, Inc.; International Business Machines; Inc., Educational Teaching Aids; Sunburst Software; Addison-Wesley Publishing Company; LogoWriter; and Ventura Software supported the project by donating materials and resources. This support enabled the teachers on the regional teams to become more comfortable and gain expertise with such materials as manipulatives, calculators, computers, and software. They were better able to share personal experiences when using these materials with their students.

Preservice Teacher Education

The Kentucky K–4 Mathematics Specialist Program had a substantial impact on the preservice preparation of K–4 teachers in Kentucky. The mathematicians and mathematics educators from each regional team incorporated many of the project's ideas and activities in their mathematics and methods courses. Ten teacher educators from private colleges in Kentucky served as outside evaluators of the seminars and attended them regularly. They received a stipend, a copy of the training manual, and a free membership in the National Council of Teachers of Mathematics. All faculty members reported that they incorporated many of the project's ideas and activities in their own methods courses. One of the mathematicians in the project received grants from the National Science Foundation and the Exxon Education Foundation to revise the two mathematics courses required of all elementary school teachers in the state. During the summers of 1992 and 1993, mathematicians from the state universities, private colleges, and community colleges met to integrate problem solving, manipulatives, and technology into these courses. They produced a curriculum for the two required mathematics courses that align with recommendations from NCTM's *Curriculum and Evaluation Standards.*

CLOSING REMARKS

The Kentucky K–4 Mathematics Specialist Program was an ambitious undertaking. Professional development for teachers of mathematics has never been attempted on a scale this large in the United States. The program was highly successful and far exceeded the expectations of its developers. The reasons for the success were many. First, the NCTM's *Curriculum and Evaluation Standards* provided a common vision and goal. The mathematics education community of Kentucky accepted this vision and sought ways to attain it as soon as possible. Second, an educational reform movement in Kentucky heightened the interest in education. The Kentucky Education Reform Act of 1990 mandated a new statewide assessment plan involving alternative assessment techniques that motivated districts and teachers to seek new curriculum and teaching ideas. Third and most important, the mathematics education community in Kentucky was willing to make sacrifices and work for the vision provided by the NCTM's *Curriculum and Evaluation Standards* and the Kentucky Education Reform Act of 1990. Without the hard work and effort of many competent professionals, this project would not have affected so many teachers and children in Kentucky.

REFERENCES

Bush, William S., Susan D. Nichols, and Jonathan Prasse, eds. *Training Manual of the Kentucky K–4 Mathematics Specialist Program.* Lexington, Ky.: University of Kentucky, 1991.

Bush, William S., Sheila Vice, and Robert K. Gilbert. "Teacher's Use of Manipulatives in Primary Mathematics in Kentucky." *Kentucky Journal for Teachers of Mathematics* 2 (Spring 1989): 5–8.

Kentucky Department of Education. *Kentucky Curriculum Framework.* Frankfort, Ky.: The Department, 1992.

———. *Kentucky Instructional Results Information System.* Frankfort, Ky.: The Department, 1991.

———. *Kentucky's Learning Goals and Valued Outcomes.* Frankfort, Ky.: The Department, 1990.

National Council of Teachers of Mathematics. *Curriculum and Evaluation Standards for School Mathematics.* Reston, Va.: The Council, 1989.

National Research Council, Mathematical Sciences Education Board. *Everybody Counts: A Report to the Nation on the Future of Mathematics Education.* Washington, D.C.: National Academy Press, 1989.

24

The Michigan Mathematics
In-Service Project

Robert A. Laing
Ruth Ann Meyer

MATHEMATICS education reform in Michigan has enjoyed a long history of cooperative efforts among the Michigan Department of Education, the Michigan Council of Teachers of Mathematics, institutions of higher education, and school districts. During the past decade this spirit of cooperation has resulted in the successful accomplishment of a number of major endeavors in K–8 school mathematics: the Michigan Middle School Mathematics Resource Teacher Project (1986); *The Michigan Essential Goals and Objectives for Mathematics Education* (1988); *An Interpretation of the Michigan Essential Goals and Objectives for Mathematics Education* (1989); New Directions in Mathematics Education in Michigan Awareness Workshops (1990); the Michigan Mathematics In-Service Project (1990); and a revision of the Michigan Educational Assessment Program (1991), designed to evaluate the accomplishment of the new objectives. The Michigan Middle School Mathematics Resource Teacher Project played a significant role in establishing a precedent for large-scale cooperative in-service efforts in mathematics and science across the state and established networks for the current effort to upgrade the curriculum and instruction of K–8 mathematics statewide.

The need for mathematics in-service programs at the middle school level was created in Michigan by a certification statute that permitted teachers to teach all subjects at the middle school level regardless of the teacher's certification level or area of specialization. This statute, since rescinded, together with the teacher surplus in the early 1980s, resulted in a middle school mathematics teaching staff dominated by out-of-field teachers (Laing and Channell 1982; Hirsch 1983). With the announcement of the Education for Economic Security Act, Title II: Competitive and Cooperative Grants program in 1985, mathematics education leaders from teacher training institutions and the state department of education decided to focus the available funds on the middle school problem.

One university was selected as the fiscal agent and submitted a cooperative-grant proposal to support the development of the in-service modules. In addition, one competitive-grant proposal was written, then adapted and submitted by each of eight colleges and universities to implement the training materials being supplied under the cooperative grant within their regions. Approximately five hundred middle school mathematics resource teachers were prepared under this project in spring 1986. Follow-up activities were conducted during the subsequent school year to extend their backgrounds and support the resource teachers' training activities within their home districts.

In 1989 the Michigan Council of Teachers of Mathematics requested the authors of this article to develop a cooperative statewide project similar to the Michigan Middle School Resource Teacher Project to serve elementary school teachers. The ensuing project, the Michigan Mathematics In-Service Project (M^2IP), was designed and implemented by a statewide project staff and included the development of in-service materials to support a two-year in-service program, the preparation of in-service leaders, and the creation of an administrative structure necessary to disseminate the project across the state. The activities of this project have been supported by grants from the Dwight D. Eisenhower Mathematics and Science Education Improvement Program, and state monies provided for professional development.

M^2IP CONTENT MODULES

The workshop materials for Phase 1 of the M^2IP project were developed by teams of mathematics educators and elementary school teachers across Michigan to support forty hours of in-service programs at each of three levels—grades K–2, grades 3–6, and grades 7–8. Materials at the first two levels are organized into four content modules entitled *Whole Numbers and Numeration, Geometry and Measurement, Fractions and Decimals,* and *Statistics and Probability.* An additional module, *Introduction and Supplement,* focuses on the new directions for the teaching of mathematics, the development of problem-solving abilities, mathematics assessment, and mathematics equity. The concepts in this latter module are reinforced throughout the other modules. In addition to the areas considered in the elementary school programs, the grades 7–8 materials include number theory, algebraic ideas, discrete mathematics, and activities involving the use of graphing calculators.

Although the curriculum materials of the M^2IP project give in-service teachers opportunities to update mathematics content through teacher explorations, the primary objectives of the project are to familiarize teachers with a model for teaching mathematics and to furnish them with sufficient opportunities to implement this teaching model with elementary and middle school children. The intent is that the model will become a characterization of their teaching behaviors in the mathematics classroom. During their first few meetings, the project staff wrestled with the task of defining a model for the teaching of mathematics that would be appropriate for developing the wide range of outcomes recommended for elementary school mathematics. Realizing that no one prescription could serve this purpose, they decided to define the model as a set of teaching principles that should be applied at every appropriate opportunity in the elementary school mathematics classroom.

The resulting set of ten mathematics teaching principles are based on research on the teaching and learning of mathematics, new priorities in mathematics education, and the teaching experiences of the project staff.

THE TEN M²IP TEACHING PRINCIPLES

Principle 1: Teachers of mathematics should view problem solving as a process that permeates the entire program and use problem solving as a context for introducing and extending concepts and skills in the mathematics classroom.

Principle 2: Teachers of mathematics should incorporate the use of children's informal real-world experiences and language, physical models, and connections within the child's mathematical background to encourage meaningful learning of mathematical concepts.

Principle 3: Teachers of mathematics should help students develop appropriate mathematical skills in accordance with the following guidelines:

*a)*The need for skills should arise from the context of problem situations or other mathematical activities.

*b)*Skills should be linked to their underlying mathematical concepts and principles.

*c)*The practice of mathematical skills should be brief and focused.

Principle 4: Teachers of mathematics should create classroom environments that foster in all students the development of mathematical power, self-esteem, and confidence in their ability to do mathematics.

Principle 5: Teachers of mathematics should help students appreciate the connections between mathematics and other disciplines as well as between mathematics and their daily lives.

Principle 6: Teachers of mathematics should furnish students with opportunities to communicate mathematical ideas in both written and oral forms.

Principle 7: Teachers of mathematics should encourage students to validate their problem solutions.

Principle 8: Teachers of mathematics should provide for periodic review.

Principle 9: Teachers of mathematics should take advantage of the power of technology to enhance mathematical learning.

Principle 10: Teachers of mathematics should engage in ongoing evaluation of learning.

These principles are implemented throughout the lessons of the M^2IP in-service programs. Participating in a model lesson, analyzing the lesson in relation to the ten mathematics teaching principles, and adapting the lesson by grade levels for implementation in the participants' classrooms prior to the next session are stressed in each session. The first activity of the following session presents an opportunity for sharing these classroom teaching experiences in grade-level groups. This structure necessitates that the workshops of the first phase of the in-service program be scheduled across an entire school year so that ample time is available for exploring these behaviors in the teacher's elementary or middle school classroom. A model lesson follows to illustrate how the format and teaching principles are reinforced in a lesson on comparing and ordering decimals for fifth- or sixth-grade students. The instructions and comments that a teacher could make to the students are given in italics.

M^2IP MODEL LESSON

Rationale: To teach the comparing and ordering of decimals in a meaningful manner

Outcomes: After completing this lesson, students will be able to compare and order decimals using concrete models, word names, or decimal symbols and solve problems involving the comparing and ordering of decimals.

Materials: Classroom sets of base-ten blocks, metersticks, and play money in sufficient quantity for each small group to have one decimal model of each type

Teaching/Learning Procedures	Commentary
Introduction: Introduce the lesson by having the students review the models for decimal place value.	As you circulate among the groups, listen to their explanations of the representations and make sure that they identify the whole within their explanations.
Use one of the models to explain the meaning of the decimal 0.65 to members of your group.	
Write the decimal on the chalkboard and then have some of the students share their representations and verbal descriptions with the whole class.	
Exploration: Students should be in small groups with a set of each of the place-value models at their tables. Present them with the following problem situation:	
Answer the question in the following situation. Use the models to explain your reasoning to your group.	Some students will object to the singularity of the word *day* in the problem statement, noting that there appears to be more than one day whose rainfall exceeded that of Sunday. Explain that this suggests that problems may have more than one solution.
Problem Situation: Assume the following table gives the amount of rainfall for the first week of July 1994. On which day did it rain more than on Sunday?	

Rainfall in Inches	
Sunday	0.26
Monday	0.3
Tuesday	0.1
Wednesday	1.7
Thursday	0.19
Friday	1.05
Saturday	0.30

Teaching/Learning Procedures	Commentary
Circulate among the groups, listening to their reasoning and posing questions to help them with verbalizing the concepts involved.	Some questions that the instructor might pose during exploration of the problem are these:
Class Discussion: To reinforce the relationships that should have been discovered in the small-group activity, pose the following questions to the class for discussion:	*What are you using for the whole? For tenths? For hundredths?*
	Which is larger, 0.3 or 0.19? Why?
Are there any days on which it rained the same amount? Demonstrate this with your materials. We call the decimals 0.30 and 0.3 equivalent. Suggest another pair of equivalent decimals. Use a model to convince your group that the two decimals are equivalent.	According to the table, it appears that there was the same amount of rainfall on Monday and Saturday. This assumes that none of the numbers was rounded. Students might use 3 longs and 30 units to show that 0.3 and 0.30 are equivalent.
Does the number of digits in a number determine which number is larger? Why? Give examples to illustrate your thinking.	A decimal such as 0.37 is greater than 0.2; however, 0.2 is greater than 0.13.
Applications: The following applications can be assigned to the small groups to reinforce the ordering and comparing of decimals and to extend the student's ability to apply the concept in the real world.	
1. Use the digits 3, 5, and 7 exactly once to write the largest number that is possible. Now write the smallest number that is possible.	1. The largest number is 753 and the smallest number is .357.
2. Place the decimals 0.3, 2.3, 0.25, 1.0, and 0.20 on the number line.	2. The order of placement on the number line is 0.20, 0.25, 0.3, 1.0, 2.3. Their approximate positions should be discussed.
3. Use a calculator to help you arrange the following fractions in decreasing order: $$\frac{5}{12} \quad \frac{2}{5} \quad \frac{4}{9} \quad \frac{3}{8}$$	3. A decimal representation for 5/12 is 0.416667, for 2/5 is 0.4, for 4/9 is 0.4444444, for 3/8 is 0.375.
4. Explain why 263 cents can be written as $2.63.	4. 263 cents is equivalent to 200 cents and 63 more cents, or 2 whole dollars and 63 hundredths of a dollar.
Evaluation of Learning Outcomes:	
1. You have the opportunity to observe students as they order and compare decimals by using concrete models, word names, and symbols throughout the lesson.	1. By posing appropriate questions during the exploration, class discussion, and application components of this lesson, the instructor can evaluate whether students are making the connections.

Teaching/Learning Procedures	Commentary

2. To give students the opportunity to synthesize their experiences with ordering decimals, to express their mathematical ideas in writing, and to furnish additional feedback regarding their perceptions, make the following assignment and collect it:

One of your classmates is absent and missed the lesson on comparing decimals. Write a note to send home to your classmate that describes how you decide which of two decimals is larger.

Dear Mary,

   To answer the question of how you can decide which of two decimals is larger using base-ten blocks, I would let the flat represent one unit, the long represent one tenth, and the small cube represent one hundredth. Then by representing each of the numbers that I wish to compare with the appropriate number of pieces, I can see which one is larger. Try it!
   Come back soon.

Your friend,
George

Analyzing the Model Lesson

After in-service teachers have experienced an M²IP model lesson, they are given the opportunity to analyze it with respect to the ten basic mathematics teaching principles. It is expected that this continual exposure to the principles will help them to apply the principles in their mathematics classrooms. The following examples from the lesson on comparing and ordering decimals illustrate how lessons are analyzed with respect to the basic principles.

Analysis of the Model Lesson	Commentary

1. Look back on the lesson and determine how problem solving permeates the entire lesson. (Principle 1)

Instead of giving the students a rule for ordering decimals and then having them practice that rule, the exploration presented a problem-solving environment in which the student was required to synthesize previous knowledge to solve the rainfall application. Application 1 also furnished a problem-solving environment while reinforcing the ordering concept.

2. How was Principle 2 on concept learning and mathematical connections applied in this activity?

3. Describe how other teaching principles were applied in the model lesson.

Physical models and the student's previous understanding regarding decimal place value were used by the student to develop meaning for the concept of ordering decimals. This forced the learner to make connections within his or her cognitive structure. The three embodiments of place value are supplied to assist with the abstraction and to promote the transfer of the concept to a variety of applications.

THE STATEWIDE M²IP NETWORK

In order to disseminate the M²IP program across Michigan, the state was partitioned into seven regions, with each being assigned an M²IP field director. Field directors were responsible for training the workshop leaders, publicizing the project, facilitating the establishment of program sites, and supervising the Phase 1 and Phase 2 programs within their regions. These field directors are mathematics educators on the faculty of teacher training institutions across the state. Since most districts are too small to support their own workshops through the Eisenhower formula grants they receive, it was necessary for the field directors to identify consortia within their regions for the delivery of the in-service programs. During the first three years of implementation, 247 year-long Phase 1 retraining programs for approximately six thousand teachers were conducted across the state.

The Michigan Elementary and Middle School Principals Association (MEMSPA) was instrumental in publicizing the new directions in mathematics education and the availability of the M²IP in-service programs. For two years the principals' association sponsored two-day regional workshops. These workshops were conducted by the state mathematics specialist and M²IP staff and were designed to familiarize administrators with the state's mathematics objectives and assessment program, with the mathematics teaching behaviors that they should observe and reinforce in the classroom, and with information needed to establish M²IP in-service programs within their regions. The *MEMSPA Letter* (1991) advertised the programs and encouraged the participation of administrator-teacher teams so that plans for mathematics in-service programs would be collaborative within the local districts. Additional publicity for the program was supplied by articles in *Mathematics in Michigan* (MCTM 1991) and numerous communications from the state mathematics specialist and several program directors in the state department of education.

M²IP PHASE 2

During the first year (Phase 1) of the M²IP program, teachers participate in activities designed to introduce NCTM's new directions in mathematics education and explore the application of these within their own classrooms. Phase 2 is designed to encourage the continuation of these explorations and to habitualize the recommended teaching behaviors to the extent that they will characterize the teacher's instructional practices. In the fall of the second year, teachers who have successfully completed Phase 1 are invited to an evening program to receive certificates awarded by the Michigan Department of Education and to explore new activities appropriate for use in their classrooms. During this session these teachers are invited to develop local networks for the following year, establish a calendar for network meetings, and, in consultation with their M²IP instructor, delineate a set of goals to be the focus of the year's activities. This plan then forms the basis of a contract between the network group and the instructor for satisfying Phase 2 requirements. Groups maintain a log of their accomplishments and share them with the entire group at the spring meeting, at which time they receive their Phase 2 certificates. Teachers are encouraged to maintain the network as a vehicle for continued growth in the teaching of K–8 mathematics.

FUNDING M²IP ACTIVITIES

Implementing of the M²IP programs requires the coordination of a variety of funding sources. Eisenhower higher education grants have been used to develop, evaluate, revise, and disseminate in-service modules; to support administrative staff; to evaluate the effects of the project regarding teacher attitudes and practice; and to partially fund follow-up activities in Phase 2. Competitive state grants earmarked for professional development in mathematics and science were obtained to support the training of in-service leaders within each region. Eisenhower local formula grants, grants from foundations and industries, and other local funds available to school districts continue to finance the M²IP in-service programs at the local level.

EFFECTS OF M²IP IN-SERVICE PROGRAMS

During the 1991–92 school year, data were collected to probe the effectiveness of the Phase 1 activities in accomplishing the goals of the project. The evaluation instruments included scales of the Likert type and open-ended questions that focused on demographics regarding professional background and current teaching activities, teachers' beliefs, their preparation to accomplish important mathematical objectives, and the amount of class time they currently assign to activities in mathematics. The average (median) teacher participating in the 1991–92 in-service programs reported fifteen years of teaching experience supported by two undergraduate courses in mathematics and six hours spent in mathematics in-service programs during the last five years.

Table 24.1 presents the assessment of the K–6 participants' backgrounds to teach mathematics and their attitudes toward the subject before and after participating in the year-long Phase 1 program. The differences suggest that the teachers consider their mathematics preparation to be better at the end of the program and that they like mathematics more as a result of the M²IP training. The paired *t*-tests used to measure these differences were significant at the 0.01 level.

Table 24.1

Pretest and Posttest Means of Teachers' Self-Assessment Regarding Mathematics Background and Feeling toward Mathematics

Item	Grades K–2 ($N = 106$)		Grades 3–6 ($N = 219$)	
	Pretest	Posttest	Pretest	Posttest
My Mathematics Background (1 = insufficient, 5 = excellent)	3.179	3.575	2.968	3.548
My Feelings about Mathematics (1 = strongly dislike, 5 = strongly like)	4.038	4.226	4.128	4.320

Figure 24.1 presents pretest and posttest data for the grades K–2 teachers' self-evaluations of their preparation to develop important mathematical objectives. Growth during the first year of the project was significant ($p < 0.01$) for all objectives. The relatively small change regarding computational skills reflects the lack of emphasis of this objective within project modules. The preparedness level on the writing objective may reflect the lack of sophistication in the use of writing by the primary school child. The continued lack of familiarity with the four-function calculator reflected a weakness in program materials corrected during later revisions of the K–2 materials and included in the Phase 2 program. Results related to an appropriate set of objectives for the grades 3–6 classrooms were similar.

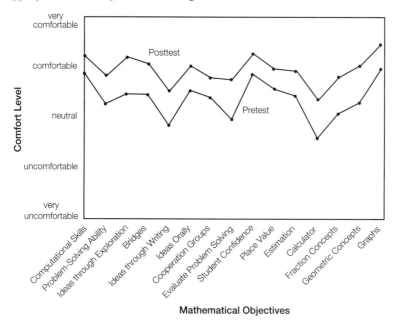

Fig. 24.1 Pretest and Posttest Means of Grades K–2 Teachers' Comfort Level Regarding Fifteen Mathematical Objectives

The evaluation instruments were also designed to reveal changes in classroom practice that occurred during the Phase 1 program. Teachers were asked to estimate the percent of a typical week's mathematics time assigned to each of eight types of classroom activities. Figure 24.2 shows the changes (posttest % minus pretest %) that these activities underwent with grades 3–6 teachers during the program. The heights of the median changes as shown on the box plots suggest that the time assigned to teacher presentations, drill and practice, and students' working alone decreased during this period, whereas the time on other classroom activities recommended for more emphasis by the new directions increased. By visualizing a horizontal line at zero on the vertical scale, one sees that 75 percent or more of the participants increased class time for using manipulatives, small-group activities, and writing about mathematics. The box plots also indicate that the time assigned to most activities changed less than 20 percent for half of the teachers. The changes of class time assigned to writing about mathematical ideas exhibits much less variability compared to the other activities, indicating that these changes were more consistent across grades 3–6 participants. The outliers at the top and the bottom of the box plots are curious and deserve further study.

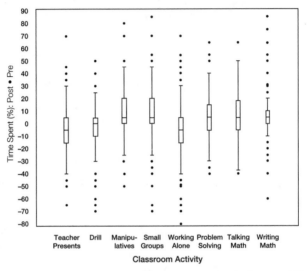

Fig. 24.2 Differences of Pretest and Posttest Responses Regarding the Percent of Class Time Grades 3–6 Teachers Assign to Eight Types of Activities

The amount of time grades 3–6 teachers spent on the eight activities appeared to change significantly as a result of their participation in M²IP and in the directions encouraged by the project. All t-tests for these items were significant at the 0.01 level.

The K–2 data indicated an increase in class time assigned to problem-solving explorations, students' talking about mathematical ideas, and students' writing about mathematical ideas. These increases were significant at the 0.01 level. Changes related to other class activities were not statistically significant in the K–2 data.

BIBLIOGRAPHY

Coburn, Terrence G., Ann Towsley, and Judy Zawojewski. "Fractions and Decimals." *Michigan Mathematics Inservice Project.* Lansing, Mich.: Michigan Department of Education, 1991.

Coxford, Arthur F., Linda Burks, and Miriam Schaefer. "Geometry and Measurement." *Michigan Mathematics Inservice Project.* Lansing, Mich.: Michigan Department of Education, 1991.

Dossey, John A., Ina V. Mullis, Mary M. Lindquist, and Donald L. Chambers. *The Mathematics Report Card: Are We Measuring Up?* Princeton, N.J.: Educational Testing Service, 1988.

Hirsch, Christian R. "Update on the Preparedness of Junior High School Mathematics Teachers in Michigan." *Mathematics in Michigan* 22 (Spring 1983): 3–5.

Kolnowski, Linda W., Ann Beyer, and Sally Roberts. "Whole Numbers and Numeration." *Michigan Mathematics Inservice Project.* Lansing, Mich.: Michigan Department of Education, 1991.

Laing, Robert A., and Dwayne E. Channell. "Teacher Reassignment and the Need for Mathematics Inservice." *Mathematics in Michigan* 21 (Spring 1982): 2-4.

Laing, Richard A., and Ruth A. Meyer. "Introduction and Supplement." *Michigan Mathematics Inservice Project.* Lansing, Mich.: Michigan Department of Education, 1991.

Mathematical Association of America Committee on the Mathematical Education of Teachers. *A Call for Change: Recommendations for the Mathematical Preparation of Teachers of Mathematics.* Washington, D.C.: The Association, 1991.

Mathematical Sciences Education Board and the National Research Council. *Counting on You: Actions Supporting Mathematics Teaching Standards.* Washington, D.C.: National Academy Press, 1991.

Michigan Council of Teachers of Mathematics. "Michigan Mathematics Inservice Project (M^2IP)." *Mathematics in Michigan* 30 (Winter 1991): 2–6.

———. *New Directions in Mathematics Education in Michigan.* Lansing, Mich.: The Council, 1989.

Michigan Elementary and Middle School Principals Association. "State of Mathematics Workshop." *MEMSPA Letter* 17 (December 1991): 8.

Michigan State Board of Education. *An Interpretation of the Michigan Essential Goals and Objectives for Mathematics Education.* Lansing, Mich.: The Council, 1989.

———. *The Michigan Essential Goals and Objectives for Mathematics Education.* Lansing, Mich.: Michigan State Board of Education, 1988.

National Council of Teachers of Mathematics. *Curriculum and Evaluation Standards for School Mathematics.* Reston, Va.: The Council, 1989.

———. *Professional Standards for Teaching Mathematics.* Reston, Va.: The Council, 1991.

Shulte, Albert P., Jo Ann K. Okey, and Richard Strausz. "Statistics and Probability." *Michigan Mathematics Inservice Project.* Lansing, Mich.: Michigan Department of Education, 1991.

25

The Pittsburgh
Mathematics Collaborative

Leslie Salmon-Cox

ISCUSSIONS on instructional change and improvement frequently focus on
D new teaching materials and new ways of preparing teachers. One of the
thornier issues in the conduct of change, however, is specifying means—and
rationale—for seasoned teachers to change their practices. Why should a professional of years standing decide to engage in new behaviors? Certainly changing
old ways can be punishing. What are the rewards? And how can such change be
facilitated by others in the environment?

As an attempt to address these issues, the Urban Mathematics Collaboratives
(UMC) program was begun in 1985 with support from the Ford Foundation.
Almost a decade later, there are collaboratives in fourteen cities across the
country: Cleveland, Ohio; Columbus, Ga.; Dayton, Ohio; Durham, N.C.; Los
Angles, Calif.; Memphis, Tenn.; Milwaukee, Wis.; New Orleans, La.; Philadelphia, Pa.; Pittsburgh, Pa.; Saint Louis, Mo.; San Diego, Calif.; San Francisco,
Calif.; and Minneapolis, Minn. Each has evolved from an initial set of objectives to later, quite distinct ones. Each has a local history, yet all have com-
monalities. This article looks at the general concept of a UMC project as a
mechanism for staff development and at aspects of the particular history of one
collaborative, exemplary of some of the more general principles.

THE UMC CONCEPT

The Urban Mathematics Collaboratives (UMC) project grew out of the Ford
Foundation's concern for the conditions that prevailed in urban schools and
classrooms. The program was conceived as a mechanism for addressing teacher
isolation, and in doing so, for improving the education of urban youth. Although
the foundation left program specifics to be worked out uniquely at each site,
certain core features for all collaboratives were mandated: all were to be set in

266

urban areas, and they were initially designed to work with high school mathematics teachers only. Each was to have an advisory board, representative of the community and including business and industry people as well as members from institutions of the higher education community. At each site, those guiding the collaborative were urged to create collegial networks for these teachers, bringing them into contact with others in the community who engage in mathematics activity, from college professors to business people of all descriptions. A guiding idea was that teachers would feel professionally enhanced by interacting with other teachers and nonteacher mathematics users and would, in fact, learn about uses and applications of mathematics that might be of interest to students in classroom activities. In an early document about the collaboratives and their objectives, there is a statement of their purpose, as viewed by the Ford Foundation (1987, pp. 7–8):

> The UMC initiative is part of the Ford Foundation's ongoing effort to strengthen institutions working to improve the economic well-being and quality of life of the urban poor. Among these institutions are community based organizations, human-services agencies, and schools. For reasons of social justice and economic health, assisting inner-city schools is vital work.... In 1950 less than 15 percent of those eighteen and under were minorities; today the figure is 36 percent and growing.

As can be seen in this statement, the strong themes in the early years of the collaboratives were *teacher enhancement* and *ameliorating the conditions in which urban youth were educated.* "Improvement" is the operative term here, since at the time the working hypothesis seemed to be that good mathematics instructional models were available and simply needed to be made accessible to these urban, isolated teachers. After several years of stressing the building of networks at each site, what became clear was the paucity of good instructional models for any teachers or students. And so the UMC joined the larger chorus of those calling for a complete reform of mathematics education.

The Ford Foundation had been specific about the need to create networks as a strategy to reduce teacher isolation. In each collaborative an early period of activity had as its goal the bringing together of various people who use mathematics in their daily lives. There were tours of numerous workplaces and exchange visits among educators and business people. In many collaboratives, problems that might serve as the basis for student lessons were jointly designed by teachers and people from a variety of backgrounds. Over time the collaboratives have differed in the degree to which maintaining these connections has been a core specification of their projects, but for each, the building of these connections was a signal initial effort.

An early component of the UMC was a documentation project designed to capture the experiences of the several individual collaboratives as well as the program as a whole. In one of the earliest reports of the documentation team is a reference to the intentions of the Ford Foundation in establishing the program, one that confirms what has been said so far about the objectives of the UMC (Webb et al. 1988, p. 1):

The teacher remains the centerpiece of the educational enterprise but—especially in inner-city schools—is likely to be overworked, lacking in support services and material resources, and isolated from other teachers, other professional adults, and changing ideas about mathematics. The collaborative project is rooted in the premise that collegiality among professional mathematicians can reduce teachers' sense of isolation, foster their professional enthusiasm, expose them to a vast array of new developments and trends in mathematics, and encourage innovation in classroom teaching.

Although the earliest focus was on the classroom teacher, and especially on providing that teacher with opportunities that would break into the normal routine of isolation and loneliness, it is interesting to note also that the history of the UMC became a history of shifting concerns and activities. Attention evolved from concerns that were teacher centered to curricula, to student outcomes and their proper assessment; from the local to the national; from a focus solely on high school teachers to more inclusionary models for teachers at all levels in many of the participating school districts. At the beginning, this evolution of concern, which marks also a degree of accomplishment of earlier goals, was unimaginable and was not foreseen by any of the major participants.

Beyond the core specifications, a principal feature of the collaboratives has been the variations among them from the start. They varied along the following dimensions:

- Where the locus of control for the project was housed
- The qualifications and background of those directing the project
- The programmatic focus, or substantive agenda, for mathematics
- The number and proportion of participating teachers, given the total population of teachers
- The degree of interpenetration of each collaborative's agenda with the agenda of the mathematics department of the collaborating school district

As noted earlier, a documentation effort was established in the first year of the UMCs. This was only the first part of a fairly elaborate infrastructure the Ford Foundation created to support and nurture the projects. In addition to the documentation materials, housed at the University of Wisconsin at Madison, a technical assistance group was created at the Education Development Center (EDC) in Newton, Massachusetts, capable of providing help to each individual project. Subsequently, EDC became the locus, as well, for the Outreach Project, designed to assist both the collaboratives as a group and others seeking to emulate the collaboratives' objectives. The existence of organized support for these three functions—documentation, technical assistance, and outreach— helped to solidify the national character of the UMC, whereas each collaborative worked hard to develop its local character and identity. The existence of the group as a whole, and the regular meetings of its participants, was a significant contribution to building the self-confidence and actual achievements of each of its individual members. The UMC started with teacher collaboration and the conditions of inner-city schools as its main foci. During the evolution of the collaboratives, mathematics issues came to the forefront nationally. In the collaborative cities, teachers and districts were poised to respond to the new challenges and have done so in varying ways.

THE NATIONAL CONTEXT

As readers of this volume are aware, American youth were shown, throughout the 1980s and into the 1990s, to be poorly served by their educational systems with respect to mathematics. International comparisons plainly showed the poor state of American youth's achievement (Lapointe 1989). Further, the least served were those quickly becoming the majority of the entering labor force, those very students about whom the Ford Foundation was most concerned (Johnston and Packer 1987). Numerous reports and calls for action have been issued (National Research Council 1989, 1990; NCTM 1989, 1991). What is clear now (but was nowhere near as obvious a decade ago) are the profound changes needed to meet the challenge of the present and the future. Fortunately, because of knowledge only recently acquired, these changes have the potential for being made. A body of research on how students learn has been conducted in the past several decades, and it is this work that, in part, serves as a basis for new instructional approaches. A growing number of successful methods for teaching, for imparting understanding—especially within the domain of mathematics—are being tried around the country. Finally, demographic projections coupled with economic analyses make clear the need: every young person, male and female and from every ethnic background, will be needed in a labor force that must be increasingly literate, both verbally and mathematically.

THE PITTSBURGH STORY

It is within the evolving UMC model and the national context for mathematics instruction that the Pittsburgh Mathematics Collaborative (PMC) has taken shape and grown in strength. The PMC's history is one of growing teacher empowerment at the very moment in mathematics education when this was most called for if reform was to take place.

Guiding Principles

Before detailing what has happened in Pittsburgh, it will be useful to sketch in several broad strokes the principles that have guided this work. These principles are five major working hypotheses derived from the sociology of the professions (Parsons 1954, Freidson 1986). First, it is important to understand the *normal* working conditions of classroom teachers. To be altered, they must first be understood. It has been well documented that teachers typically work in isolation, lack collegial relationships, and tend to have occasionally difficult relationships with school-based and other administrators (Lortie 1975; Miles 1980; Fullan 1991). It was essential to break into this isolation and create opportunities for professional and social interaction (Driscoll and Lord 1990). To enhance professional development, it was essential to go beyond social occasions, to stress task orientation. Working on tasks reveals a teacher's expertise as well as places where further development is needed. Pittsburgh's Mathematics Collaborative has relied strongly on the hypothesis that *when recognized, knowledge can be transformed into authority.*

That hypothesis is part of a second notion, namely, that *as teachers recognize and value their own expertise, they become more effective.* At the core of a profession is its claim to a knowledge base that cannot be shared by lay publics. Not only do secondary school mathematics teachers have such a base, but in the United States, at least, it is increasingly rare (that is, in many urban areas of coastal America, those teachers assigned to mathematics classes are not necessarily certified to teach mathematics). Their knowledge of mathematics and of pedagogy, then, makes these teachers a scarce resource to be valued and nurtured. This view was novel to the UMC teachers, but they have come to appreciate it.

A third hypothesis is that the *acknowledgment of teachers' expertise by others, especially by school administrators, improves performance.* There is an unfortunate history of decision making in education. In many school districts, educational decisions are made by bureaucrats, at worst lacking the requisite knowledge and at best lacking current knowledge. Similarly, state legislatures frequently mandate educational policy. For education to become a profession—parallel to those of law and medicine and capable of generating the kinds of reforms that are being called for—teachers' professional expertise must be recognized and made use of. Management is in the province of the bureaucracy; issues concerning the learning of subject matter are for teachers and properly trained supervisory staff.

Fourth, it was clear at the outset that *creating a structure—in this situation through the collaborative—that would increase opportunities for teachers to exercise authority could result in improved performance.* Probably, at least initially in such an empowerment effort, this structure needs to be outside the school district it wishes to affect. This is perhaps one of the clearest insights that has been confirmed over time. There has been a great deal of discussion about "restructuring schools." The Pittsburgh experience suggests that some of the teachers were more than ready to embark on new, profession-enhancing activities, even as the schools they worked in were showing few signs of change. The collaborative—and its initiatives—was an alternative and different normative structure within which teachers could act according to a new model of themselves. Simultaneously, structural change was under way in Pittsburgh. As Pittsburgh schools continued to be restructured, the newly enhanced mathematics teachers were ready to be leaders in them. That restructuring leads to the next point. For changes to occur—at the building, departmental, or district level—there must be systemic flexibility. Change is impossible in the face of rigidity.

Finally, there is the issue of standards, which are the hallmark of a profession. *Articulated standards for improved mathematics curricula* are essential to a project such as the UMC. Fortunately, the National Council of Teachers of Mathematics (NCTM) began the process of establishing standards as the collaborative was taking shape. The implementation of these Standards, and especially their acceptance by teachers, became a major goal of the collaborative. It became clear that if Pittsburgh's curricula came to reflect the NCTM Standards, then courses of study would truly reflect the new reform effort. This would enhance both the teachers' self-concepts and their actual behaviors.

The PMC

The Pittsburgh collaborative was initiated in fall 1985 and was staffed with a coordinator and an assistant coordinator. (At that time, the Pittsburgh Public Schools was reorganizing and a search was on for a director of mathematics, a position eventually filled six months after the start of the collaborative.) An initial strategy was to create a group composed of the chairs of the mathematics departments. This strategy is directly related to the guiding principles enumerated above. It was designed to reduce isolation and enhance collegiality. It was a mechanism for recognizing teachers' knowledge. By arranging for these teachers to interact with administrators, their knowledge base was recognized and accorded the respect it deserved. The group, and over time the quality of its meetings, departed from then-standard operating procedures and allowed its members to behave in new ways.

Creating a group involves a great deal more than simply convening it. When the collaborative began, there was a chair in each mathematics department. These people never met as a group. Their identification as chairs meant simply that they had a few more responsibilities for a very slight increase in salary. They were appointed to their position by their principal. Some principals consulted their faculty; some did not. The collaborative initially convened these people for a dinner and a speaker followed by a half-day workshop focusing on needs assessments.

The chairs expressed a strong interest in several opportunities the collaborative seemed able to offer, including learning more about mathematics applications; offering more and better options for curricular change, particularly for the non-college-bound students; and—most especially—becoming more involved in the decision making that affected their daily lives as mathematics teachers. On this last point, however, they were extremely skeptical regarding the likelihood that they would be included in a decision-making capacity. These teachers' past experience led them to believe that even if they were to organize around a substantive agenda for action, their involvement would not be welcomed by the administration of the district.

An early issue the collaborative addressed was the newly legislated state mandate for three years of high school mathematics. Prior to that time, only two years were required. The teachers were quite concerned about who would decide the content of the new course. They were convinced that this would be an administrative decision, made without consulting them. The collaborative coordinator met with the superintendent of schools and convinced him that the chairs were the natural and obvious group to design the content of the new course. Therefore the collaborative was launched, having arranged for a decision that acknowledged teachers' expertise, one the teachers had been convinced could not possibly be made to their satisfaction.

Within six months, the district had filled the director of mathematics position with a young, highly qualified mathematics educator willing to share responsibility and power. The chair group, created by the collaborative coordinator, became the "Mathematics Curriculum and Policy Advisory Group," a designation given them by the newly appointed director. The creation of this group was crucial for the identity of the collaborative. Meetings were task oriented and centered on substantive issues. Early issues included the new third-year course, the development of a guide for counselors to place students in appropriate classes, and the devising and implementation of strategies for using calculators in all mathematics classes. The director treated these teachers as the subject-matter experts they were. Increasingly, they recognized the power of their expertise. Meeting regularly with the director and her supervisory staff—heretofore never done—gave this group an identity as a group, with access to power and possessing legitimate authority. So, during the first year, the group evolved through a series of stages: from alienation and complaint about administrators and working conditions to substantive dialogue about instructional matters. The teachers met on their own time throughout the year and—with two exceptions—were not paid for those meetings. (In the subsequent years, no meeting times have been paid for, and the meetings themselves have actually become longer and more time consuming.)

Since that time, the chair group has become fundamental to the structure and to decision making about issues in secondary school mathematics education in Pittsburgh. Often, initiatives that grow from discussions of this group become the agenda for a newly constituted, different group of teachers. In this way, decision making has become more shared and more teachers have become involved.

In the years since the initiation of the project, it has been expanded in significant ways. Funding from the National Science Foundation provided for the inclusion of a middle school program. Enthusiasm and persistence on the part of teachers have resulted in new courses of studies, new pedagogies, and the integration of new technologies. It would be impossible to detail all the initiatives the collaborative has undertaken in the ensuing years. Perhaps applicable to each has been an explicit ongoing principle: any activity undertaken, any new group formed, was from its initiation embedded in school-district structures. Strong and continuing attention has been paid to issues of institutionalization. The Pittsburgh strategy has been to interlock all initiatives with the school district. Coordination of the collaborative has increasingly involved the district's director of mathematics so that collaborative "routines" are district "routines"; hence a culture for teaching and learning has been broadly established.

In Pittsburgh, as elsewhere, a great deal of early activity had to do with creating interactive settings for teachers and others who use mathematics, either in business or in industry or within the domain of higher education. Connections to the business world have been continued and institutionalized by the mechanism of having the assistant coordinator be a shared employee with the local Partnership in Education program. This program's raison d'être is the building of business partnerships with schools. A major connection was established with the Carnegie Science Center. An emphasis in Pittsburgh has been on the creation of teacher networks with ties to others becoming secondary, albeit important, over time.

The history of one Pittsburgh group, other than the mathematics chairs, will be used as an example of the strategies employed, the objectives sought, and the gains won. That group is the computer group, a collection of ten individuals initially, some of whom knew how to use computers and most of whom did not. The collaborative coordinator obtained extra funding to bring these teachers together on a regular basis, to expose them to instruction in the use of computers in mathematics classes, and to build, within them as individuals and among them as a group, a resource for the entire district. Each participating teacher who needed one was given a computer to take home. Each was paid for work at home until such work was no longer viewed, by them, as separate from ongoing responsibility and remuneration was refused. As they became engaged in thinking about computer use, one of their first realizations was the paucity of good instructional software. They began to create critical filters through which they judged the materials they were viewing. Several of the teachers went to national conferences where they learned what others were doing. They began to see themselves as part of a national movement with colleagues in school districts across the country.

Locally, by the third year of the project, they were each the experts in their own buildings, offering assistance and counsel to other teachers beginning to use computers. As a group, they offered an in-service course for their fellow teachers. Finally, it was possible to link the schools to one another through the installation of modems attached to the computers in the mathematics department offices and to link all schools with a country-wide forum.

The history of this group shows that it was involved in enhancing knowledge, building a critical frame of reference, developing new collegial relationships both locally and nationally, engaging in activities that heighten a sense of professionalism, and gaining recognition for professional work. These, in toto, were the objectives of the collaborative.

Emphasis has shifted from a focus on improvement to one on reform. This is true whether with reference to teacher involvement or curricular change. Pittsburgh has moved from simply increasing teacher involvement to actual teacher leadership and the initiation of ideas. In the curricular areas, profound changes have been undertaken, since Pittsburgh has sought to align its program with that recommended by the NCTM. It is no longer sufficient simply to look for up-to-date textbooks. Now, calculators and computers must find their appropriate place along with new books, new forms of assessment, and teachers trained in new forms of pedagogy.

Several years ago Pittsburgh teachers engaged in a grants competition and won, obtaining funding for a conference on teacher leadership. The conference was planned entirely by the teachers, who hired an outside facilitator to run it. The focus of the conference was on helping participants develop communication and leadership skills. Pittsburgh teachers of the mid-1980s would not have felt confident enough to engage in such a competition. They probably would not have even realized the need for such a conference, or if realizing it, they would not have felt it possible to plan.

Finally, and in part as an outgrowth of this conference, each high school's mathematics faculty is developing or implementing plans for improvement. Each plan addresses a problem the faculty believes is a central one for its students, and each plan involves either all or most of the building's mathematics teaching staff. The development of these plans—the initiation of the idea—is undoubtedly a result of a mixture of factors. Intelligent administration of the division of mathematics is no small part of what has happened. However, a large part of the explanation is the enhanced professionalism of the involved teachers. That professionalism is largely a result of the collaborative project. Although the initial goals were centered on teacher professionalism, it is encouraging to see that later initiatives have actually affected the lives of young people. This change occurred because engaged, professional teachers decided to make a difference in student learning, which is, after all, an ultimate objective of any reform effort.

These experienced teachers engaged in all this activity because their professionalism was recognized and enhanced. They were given opportunities to exercise authority through strategic decision making. The collaborative achieved its initial goals—reducing isolation and increasing collegiality—and later surpassed them by achieving other specified goals as well.

REFERENCES

Driscoll, Mark, and Brian Lord. "Professionals in a Changing Profession." In *Teaching and Learning Mathematics in the 1990s,* 1990 Yearbook of the National Council of Teachers of Mathematics, edited by Thomas J. Cooney, pp. 237–45. Reston, Va.: The Council, 1990.

Ford Foundation. ...*And Gladly Teach*.... New York: The Foundation, 1987.

Freidson, Eliot. *Professional Powers*. Chicago: University of Chicago Press, 1986.

Fullan, Michael. *The New Meaning of Educational Change*. New York: Teachers College Press, 1991.

Johnston, William B., and Arnold E. Packer, eds. *Workforce 2000: Work and Workers for the Twenty-first Century*. Indianapolis: Hudson Institute, 1987.

Lapointe, A. E., N. A. Mean, and G. W. Phillips. *A World of Differences: An International Assessment of Mathematics and Science*. Princeton, N.J.: Educational Testing Service, 1989.

Lortie, Dan C. *Schoolteacher: A Sociological Study*. Chicago: University of Chicago Press, 1975.

Miles, Mathew B. *Common Properties of Schools in Context: The Backdrop for Knowledge Utilization and "School Improvement."* New York: Center for Policy Research, 1980.

National Council of Teachers of Mathematics. *Curriculum and Evaluation Standards for School Mathematics*. Reston, Va.: The Council, 1989.

————. *Professional Standards for Teaching Mathematics*. Reston, Va.: The Council, 1991.

National Research Council, Mathematical Sciences Education Board. *Everybody Counts: A Report to the National on the Future of Mathematics Education*. Washington, D.C.: National Academy Press, 1989.

————. *Reshaping School Mathematics*. Washington D.C.: National Academy Press, 1990.

Parsons, Talcott. "The Professions and Social Structure: A Sociologist Looks at the Legal Profession." In *Essays in Sociological Theory,* edited by Talcott Parsons. Glencoe, Ill.: Free Press, 1954.

Webb, Norman L., Susan D. Pittelman, Thomas A. Romberg, Allan J. Pitman, and Steven R. Williams. *The Urban Mathematics Collaborative Project: Report to the Ford Foundation on the 1986–1987 School Year*. Madison, Wis.: Wisconsin Center for Education Research, 1988.

26

Industry Internships and Professional Development

Ann M. Farrell

TEACHERS who spend a summer working with chemists, engineers, environmental consultants, loan officers, marketing analysts, materials scientists, and physicists learn how valuable problem solving, teamwork, and communication skills are in business and industry. They find real-world applications of mathematics to motivate students. They learn to use computers for word processing, spreadsheets, and databases. These teachers feel rejuvenated by the opportunity to try something new, to learn about careers in industry, and to collaborate with professionals from both inside and outside education.

DEVELOPMENT OF AN INTERNSHIP PROGRAM

Project GEMMA, Growth in Education through a Mathematical Mentorship Alliance, is coordinated by the Alliance for Education. The alliance brings together teachers and administrators from area school districts, mathematics and science educators from two universities and a community college, and scientists and executives from several area industries and businesses. Participants plan and take part in a variety of activities, with the purpose of improving the mathematics education of all students in the county's public schools.

The objectives of Project GEMMA include the following:

1. Helping teachers discover connections between mathematics and science, engineering, and other disciplines used in the workplace
2. Disseminating to teachers materials about current real-world applications of mathematics
3. Providing students with examples of mathematical applications that require problem-solving skills and that they will encounter in their life after school
4. Forming a network between mentors and teachers, mentors and other mentors, and teachers and other teachers

5. Empowering a group of teachers to become leaders through a unique learning experience and reinforcing their leadership skills with opportunities to interact with other professionals at local, state, and national meetings

6. Allowing teachers and mentors to examine how the curriculum and instruction that follow the guidelines set forth in the NCTM's (1989) *Curriculum and Evaluation Standards for School Mathematics* can improve students' preparation for careers in business, science, and industry

THE IMPLEMENTATION OF PROJECT GEMMA

The program was initiated with ten teachers participating during the "pilot" summer of 1990. Eighteen teachers took part during summer 1991, twelve teachers during summer 1992, and twenty-three during summer 1993; twenty-nine will participate in 1994.

Summer internships begin in June and run for either eight or ten weeks, with most teachers choosing the eight-week option. The sites that have hosted or are hosting teachers include the consumer-marketing office of a bank (Bank One, NA), three scientific consulting firms (Woolpert Consultants; QSource Engineering, Inc.; and The Analytic Sciences Corporation [TASC]), two materials-science laboratories (Wright-Patterson Air Force Base Materials Directorate and EG&G Mound Applied Technologies), two divisions of an automobile manufacturing company (Delco Chassis Division and Harrison Division, both of General Motors Corporation), the research and development arm of a copy-machine producer (Scitex Digital Printing, Inc., formerly a division of Eastman Kodak Company), two divisions of a paper-production and information-services company (Mead Corporation and Mead Data Central, Inc.), and a hospital (Children's Medical Center).

Each teacher is assigned to a mentor who guides him or her through the completion of a project at the industry or business. The teacher is expected to make a contribution to the industry, not be just a spectator. At the same time, teachers are observing how people in industry operate, what skills they need, and, in particular, how they use mathematics in their daily work. Examples of the projects teachers have worked on include analyzing marketing surveys for the consumer division of a bank, testing fan blades for automobile engines, analyzing statistical data from automobile dealerships about the maintenance and failure of automobile parts, researching the operation of a microwave that is ready to be installed on a factory production line, researching the environmental impact of building construction, gathering data from computer simulations of metal extrusion processes, determining a formula for describing the path a ball bearing takes around a rim, studying and testing new polymers, determining and graphically displaying the relationship among molecules in a new material, creating a computerized model of transportation systems, and testing ink jets in photocopiers to improve their specifications.

During the internship, the teachers attend four seminars, which are facilitated by mathematics and science educators from the area universities and community college. Each afternoon's agenda includes allotted time for teachers to discuss their experiences in the workplace, the applications they are learning about and intend to share with their students, and the teaching methods they are considering as a result of what they observe in industry. The seminars are held at the industry sites involved in the program. Often teachers take a short tour, and interns who work at that site describe and show their projects.

At the end of the summer, a luncheon is held at the Engineers Club of Dayton. Teachers give short presentations about their summer experience to past and present participants in Project GEMMA, school administrators, university mathematics and science professors, and representatives from businesses and industries who are hosting or are considering hosting teacher-interns.

During the academic year following their summer internship, the teachers pilot the applications problems they and their colleagues have written. The goal is to develop a booklet of applications problems that have been written, piloted, and revised by project participants. Mathematics workshops are planned during which Project GEMMA teachers can share with other teachers from around the region these problems and plans for incorporating them into the curriculum. These problems show students how mathematics is used in local businesses and industries.

In two follow-up "reunions" during the academic year, the teacher-interns can reflect on what they experienced the previous summer and what they are trying to transfer to the classroom for their students. We invite participants from past summers to the reunions as well so that the teachers can widen their circle of colleagues and stay involved from year to year. The teachers share their triumphs and their frustrations and learn from one another how to transform successful industry content and methods into successful classroom content and methods.

PLANNING FOR THE INDUSTRY INTERNSHIPS

Preparations begin every fall for the following summer. By the time teachers see advertisements and applications for the program in January, the staff at the Mathematics Collaborative has already been communicating with area industries and businesses to plan their participation. A representative from each industry site completes a "preferred skills form" listing the mathematics, science, and computer background necessary for a teacher to be comfortable in a placement at their site. Most sites choose more than one project that would be appropriate for a teacher-intern and refine their plans after they meet the teacher chosen for their site.

The teacher-applicants are interviewed by a panel made up of one or two industry representatives, a past teacher-participant, and the coordinator of placements. Interviews take place on several afternoons and one Saturday in March. During the interviews more is learned about teachers' backgrounds, their teaching assignments, their enthusiasm for the program, and what they hope to gain from the experience. The panel answers any questions the applicants have about the program. The coordinator of placements, a mathematics educator, sits in on every interview. She has read all the industry preferred skills forms and all the teacher applications so that she knows the placements for which each teacher appears to be qualified. By listening to the questions that the industry representatives ask during the interview process, she can clarify further the type of background needed. After all applicants have been interviewed, the panel finds the best match between applicants and available positions.

The industries desire more computer skills than most of the teacher-applicants have, but they have always rearranged their plans so that the teachers can have time to learn these skills. Most teachers are able to compensate for what they lack in specific mathematical or technical knowledge by a willingness to learn and an enthusiasm for the opportunity to work in industry. The project's directors have been pleased with the flexibility that both teachers and mentors have shown in making the experience positive for everyone.

After the assignments have been made, a "prementorship orientation" is held so that teachers and mentors can meet one another before the teacher arrives at the business. Expectations and plans for the summer are discussed. Some anxiety is felt within both groups, which is similar to the anxiety felt by any new employee or employer. To reduce the uncertainty for the teachers, the mentors share information about policies and practices at their site. They share such information as workday starting and ending times; arrangements for parking; facilities for meals; acceptable (and in the example of some manufacturing plants and laboratories, practical) dress; medical policies (for example, what type of physical examination is necessary to get on the payroll); safety codes; names, titles, and telephone extensions of persons with whom the interns will be in contact; what interns can expect on their first day; and any other information pertinent to a new employee. To introduce the mentors to the teachers' world, we ask the teachers to arrange a classroom visit for their mentors and share lists of objectives for the courses they teach. Those course objectives and a wish list of topics for which the teacher would like to find applications to show their students help mentors tailor relevant activities for their teacher-interns.

EVALUATING THE INTERNSHIP PROGRAM

The Alliance conducts both formal and informal evaluations of the program and its components. The formal evaluations consist of questionnaires completed by the teacher-interns near the end of the internship and again during the academic years following the internship, questionnaires completed by the mentors at the end of the internship, debriefing sessions held separately for teachers and mentors at the conclusion of the internship, and observations reported by an industry representative who attends the summer and academic-year seminars.

Our informal evaluation of the program includes open discussions held during the seminars. The comments teachers share with us then help us to plan the next seminar and alert us any problems at the sites, as well as help us make adjustments in our procedures for the following summer.

One informal measure of the value of the program is the retention rate among our industry partners even during economically difficult times. Our industry hosts have noted that they benefit from working closely with educators and learning more about mathematics curricula, teaching, and students through their teacher-interns.

Almost all the teacher-interns say they would participate in the program again if the opportunity was available. Those who say they would not remark that their participation does not allow them enough vacation time or that they don't think that repeating the experience would be meaningful. Eight teachers have returned for a second summer and all report benefits from participating for a second summer. These teachers help new participants and mentors become acclimated to the program more quickly.

The aspects of the program that teachers like or enjoy most fall into four categories: (1) enjoyment of the actual work they perform in industry, (2) the chance to learn something new, (3) the interaction with engineers and scientists in industry, and (4) the interaction with other educators. The evaluators have been pleasantly surprised by the variety of professional growth opportunities that teachers find inherent in the program, as indicated by their responses to the end-of-summer surveys.

The teachers' comments about the work they perform indicate that they enjoy the "hands-on work"; the "lab work"; and the "opportunity to be part of a team and actually participate in making a product," not just tour and observe. Teachers say they "enjoyed the day-to-day experience of walking in another set of shoes." They enjoy the "general exposure to something new," "the variety of the projects," "learning more about computers and technology," and "working in something I knew nothing about." These comments should not be surprising since we would expect teachers to enjoy learning!

Many of the teachers say that becoming a learner again makes them better teachers. They note the importance of their mentor's attitude toward them whenever they encounter difficulties in learning something new on the job. They report being more understanding of students' frustrations.

The following remarks indicate that the teachers appreciate their interactions with engineers and scientists: "I enjoyed being in the 'real world' with a chance to see what other professionals are doing." "I enjoyed meeting people who use the math I teach on a daily basis—sharing ideas with them, learning from them, seeing technology in action." "I had many discussions with all kinds of engineers about their projects and their attitudes." "It was great to work with a group of engineers who focus on analyzing problems." "I enjoyed the interaction with so many people concerned with improving education."

Between January and March 1992 (after either one semester or one year plus a semester following the summer internship), a questionnaire was completed by twenty-one of the twenty-eight teachers who had participated in 1990 and 1991. Their responses to the question "What is the most important thing you feel you gained from your summer experience?" show a balance among mathematics applications, career information, confidence, and new opportunities. The first two responses were expected. The third, confidence, includes "confidence that we are teaching important, life-long math skills," "confidence knowing there are connections between the math in the classroom and practical applications," and "a confidence that I can speak with authority when I tell my students how certain math topics are used in the real world." The "new opportunities" include opportunities "to interact with professionals in my academic field"; "to use the latest technology"; to have "time at work to think, organize, and plan"; and to work cooperatively on a project.

To an open-ended question about how their instruction has improved as a result of GEMMA, the teachers respond that they give the children more time to formulate an answer, they offer more opportunities for practical application and group interaction, they use examples from industry to relate real-world situations to topics in mathematics, they are teaching with greater purpose and are more aware of the NCTM Standards, and they use cooperative learning more frequently. In drawing on their GEMMA experience to emphasize each of several practices listed, fifteen of the twenty-one respondents frequently emphasize work ethics and responsibilities to the group, fourteen frequently incorporate more teamwork on assignments, thirteen frequently emphasize interpersonal skills, twelve frequently emphasize communication skills, and ten frequently integrate both mathematics and science concepts into their curriculum. These self-reports form a basis for the more formal classroom observations that will be conducted we plan to do to assess the impact of the internships on students.

The evaluations have revealed the following about the program's components:

Seminars. The teachers value the time they have to talk with other participants about the experience and its implications for the classroom. They want ample time for small-group discussions. Each summer, several teachers have earned university credit or continuing-education credit from the state for the work they do at the seminars.

Several science teachers involved in an industry mentorship program funded by the U.S. Department of Energy (the Teacher Research Associates program) have attended our seminars, thus giving the mathematics teachers an opportunity to confer with science teachers about the science content involved in the industry projects. The teachers have exchanged ideas about integrating mathematics and science and aligning the high school mathematics and science curricula so that, for example, chemistry students will have studied the mathematics required for a particular topic before they study the topic in chemistry. Also, the teachers have been discussing the similarities between reform in mathematics education and reform in science education and how they can support efforts to improve the teaching of one another's discipline.

We have begun to hold special sessions for the mentors at the first and third seminars. One of our industry colleagues, chosen because he understands fully the nature and purpose of the program and has been an exemplary mentor for two summers, facilitates these sessions. The mentors ask questions and raise issues about their roles if they judge that problems are occurring at their sites. Other mentors offer suggestions. Every placement is unique because of the nature of the work and because of the personal interactions between mentor and teacher; yet by listening to other mentors describe the projects they have chosen for their teachers and the particulars of which staff meetings teachers attend, what processes teachers observe onsite, what personnel they meet, and when they are expected to give presentations and to whom, the mentors are able to define further the internship experience for their own teacher-interns.

The seminars often conflict with meetings at the industry site, which poses a problem for the teachers and mentors. The mentors' presence and input at the seminars are valuable, but participation requires a special commitment because of the time away from their jobs.

Placements. Having the teacher preview the site and meet the mentor at an off-site orientation meeting makes the teacher feel more comfortable on the first day of the summer internship. It is valuable for the mentor to visit the teacher's classroom before the summer experience. The mentor sees the intern working in the school environment, where he or she is comfortable and in control. If the intern becomes intimidated by, and uncomfortable with, learning a new job in industry, both the mentor and the teacher will realize that the newness of the site and the experience is responsible.

Pairing teachers at an industry site has advantages. They benefit from daily conversations about what they are doing and observing and from the potential for transfer to the classroom.

Lacking the necessary computer skills to jump into their projects, most teachers spend the first week of their internship learning to operate the software and hardware. Ways are being investiagted to offer evening or Saturday classes in word processing and spreadsheet- and database-applications software for future summer interns so that they will be better prepared to work in industry (and in the classroom). The project may be able to tap resources available through the Alliance, through community adult education providers, or through existing university classes. The classes should be designed for the teachers' needs but use software popular in industry.

Writing applications problems. The teachers consider writing applications problems valuable for achieving the goal of classroom transfer, but it is difficult to gather feedback, finalize problems, and keep the teachers networking. Teachers need to talk to other teachers about the applications problems before they feel comfortable introducing them to their students. We believe that an academic-year class for credit may help, as would access to electronic networking facilities.

Academic-year programs, such as those developed by the Industry Initiatives for Science and Math Education (IISME) in California, are attractive to our teachers. (Project GEMMA thanks Karin Rosman, education coordinator for IISME at the Lawrence Hall of Science, University of California, Berkeley, and her colleagues who have shared information about their program with the project.) The teachers and mentors at IISME have formed an "academy" that offers workshops on topics in advanced mathematics, classroom methods, industry methods, and any other topics desired by teachers who are examining how to transfer the industry experience into their schools. The teachers have voiced an interest in workshops on cooperative learning and teaching problem-solving techniques.

Teachers who have in internships through IISME and are motivated by their experience in industry choose a theme, such as "teamwork" or "writing across the curriculum," and write lesson plans to develop it in the classroom. Project GEMMA plans to try this idea to see if the development of themes can augment the integration of real-world applications into the day-to-day curriculum.

EFFECTING CHANGE IN THE CLASSROOM

In final papers submitted at the end of the summer experience, the teachers described how the experience in industry will affect their teaching. Several participant mentioned a need to integrate mathematics with other subject areas. They saw the need for every student to know mathematics and to know how to apply it.

> It is also important to describe science and math used by occupations that are not usually considered scientific or mathematical in nature. For example, a secretary may be asked to compile a report including information about a list of chemicals and their percentages found in a sample. Math and science are not extras; they are essential for all walks of life. (David Lindamood, Patterson Career Center)

Several teachers intended to set a classroom objective to require students to work with others on problems.

> I remember most the time that a senior engineer, a student, and the two of us [Project GEMMA teachers] deliberated the algorithms we had each independently proposed for calculating the sampling rate required for a wave analysis program. We all had errors in our procedures, but by talking them out with each other we were able to come up with one algorithm that was right. None of the original suggestions was absolutely correct, but cooperation led to a good solution. What a powerful lesson for all of us to learn! I am sure that I shall use cooperative learning methods in the classroom much more than I have in the past. (Richard Vilhauer, Oakwood Senior High School)

Another teacher mentioned the difficulty that arises because the behaviors that are the most difficult to assess in the classroom are the most important ones in industry. Yet she is determined to make them objectives for her upper elementary school students.

> Reasoning, planning, self-direction, listening, communication, motivation, and enthusiasm are the backbone and support for the mathematical and scientific skills that are used by the workers in this industry. This has had a tremendous impact on the direction mathematics education will take in my classroom.... I have witnessed firsthand that problem solving is a developmental skill best achieved by working together in groups and that computers and technology cannot be over-emphasized as they are the mainstay of industry. Greater responsibility for learning will be placed upon the students. No longer should the classroom be a place of recipe following and independent regurgitation of factual information. The real world of work does not offer opportunities for these isolated functions. (Brenda Cook, H. V. Bear Elementary School)

The Project GEMMA Teachers returned to the classroom with a renewed sense of commitment to teaching mathematics.

> First, this experience has taken my enthusiasm for teaching math to a new high. I am more excited than I ever was about the benefits of a good math background. Also, I have greater insight as to why I teach math. This should give me more credibility with my students. Next, I'll be able to tie together all of my lessons with some real life applications after everything I've seen this summer. (Peggy O'Brien, Mad River Middle School)

> Personally, this experience has been good for me. I've had an opportunity to follow a "road less traveled" and find fulfillment in a field in which I'm trying to direct my students. I've felt a renewed sense of professionalism that substantiates the importance of the math I've been teaching. On more than one occasion, engineers I met told me it was a math or a science teacher that initiated their interest or gave them the confidence to pursue their career. (Tim Voegeli, Kettering Fairmont High School)

BEGINNING AN INDUSTRY INTERNSHIP PROGRAM

High-level executives from several local companies supported the mentorship project and volunteered their own companies as sites for the 1990 pilot summer internships. Without their interest and support, the project would never have got off the ground. One person in particular, a vice-president in a company whose international headquarters are in Dayton, was instrumental. He set up the initial meetings with representatives from other companies and encouraged their involvement in the program. Without such a catalyst, the project would have been much more difficult to begin.

The funding for Project GEMMA includes contributions from participating businesses for teachers' stipends, workers' compensation, and seminar costs (i.e., for facilities, lunches, and materials). The Alliance for Education provides a director, office space, secretarial support, and staff for telephone calling, copying services, and setting up meetings. The Alliance for Education is sponsored by local businesses. Project GEMMA was initiated and is maintained through the alliance's already established connections with local industries.

SUMMARY

The teachers have gained much more than we ever expected from Project GEMMA. The project directors expected them to discover applications for the mathematics that they teach. But they are also finding examples that reinforce current reforms in pedagogy, including accessible technology, cooperative learning, writing to learn, and problem solving. It was hoped that they would gain a renewed sense of professionalism through their collaboration with one another and with their mentors. This hope has been realized, as seen in the analogy drawn by one teacher after his second summer in the program:

> I also see the role that we [teachers] play in our "industry" as it relates to other industries. My position this summer was not unlike a sales representative of a company named Education, Inc. As a representative of this company, I have gone out into another industry to survey what their needs will be so that I can design a "product" that they can use. I must take pride in the quality of my "product" and be able to stand behind it.... The more input I can get from my customers, the better the "product" will be. (Jeff Weaver, Patterson Career Center)

REFERENCE

National Council of Teachers of Mathematics. *Curriculum and Evaluation Standards for School Mathematics*. Reston, Va.: The Council, 1989.

27

Using Knowledge of Children's Thinking to Change Teaching

Donald L. Chambers
Judith Elaine Hankes

PROFESSIONAL development programs can bring about significant changes in classroom practice by helping teachers make informed decisions instead of training them to perform in a specified way. Primary-grade teachers in a Wisconsin school district dramatically changed their vision of teaching and their classroom practice as a result of their participation in a professional development program.

- "I know more about how my students learn. I never really asked them before. I told them what to do." (Ms. R, a second-grade teacher)
- "I ask children how they got the answer. I don't think I ever did that before." (Ms. M, a first-grade teacher)
- "I didn't give the children enough credit for the knowledge they brought to school." (Ms. J, a first-grade teacher)

In this article we describe the philosophy of the Cognitively Guided Instruction (CGI) program, the characteristics of CGI classrooms, the CGI professional development program, and the role of the school district in offering this professional development opportunity to its teachers. Then we discuss the changes in teaching that resulted, described by the teachers themselves.

THE PHILOSOPHY OF COGNITIVELY GUIDED INSTRUCTION

The approach taken in the Cognitively Guided Instruction program is to help teachers understand children's thinking, give the teachers an opportunity to use this

The authors are grateful to Elizabeth Fennema for her valuable comments on early drafts of this article. The Cognitively Guided Instruction project is directed by Elizabeth Fennema and Thomas P. Carpenter, University of Wisconsin at Madison. It is supported in part by a grant from the National Science Foundation (MDR-8955346).

knowledge in their classrooms, and give them time to reflect on what happens as a result of using this knowledge. The participating teachers typically modify their own instructional methodologies on the basis of their knowledge of their students' understandings. A previous study (Carpenter and Fennema 1992, p. 462) concluded that

> we can change teachers by helping them to understand children's thinking and that those changes are reflected both in what the teachers do in the classroom and in their students' learning.

CHARACTERISTICS OF CGI CLASSROOMS

Although instructional practices are not prescribed in the Cognitively Guided Instruction project, teachers use their knowledge of children's thinking in the following six ways:

1. They base their curriculum on problem solving. Problem solving is both the major goal of instruction and the principal instructional activity.

2. They pose problems that are based on the child's understanding. The context of each problem is familiar to the students, and students are able to call on their previous experiences to solve the problems even though they may not be able to write number sentences, recall number facts, or perform computational procedures.

3. They encourage children to talk about their problem solutions. Different children solve problems in different ways. As children listen to strategies used by others, they have an opportunity to see if they understand strategies that are different from their own. They may incorporate another strategy into their own repertoire if they understand it and if it seems better than theirs. This communication of strategies helps children articulate their mathematical ideas, it informs other students and gives them new options, and it informs the teacher about the thinking of individual students.

4. They expect multiple strategies to be used. Seeking new strategies stretches children to find new ways to think about mathematics, which helps them increase their understanding. Furthermore, since not all students finish a problem at the same time, one way for students to use their time when they finish early is to seek other methods for finding an answer. After creating alternative solutions, they can select the method they like best. Since the teacher may, from time to time, them give opportunities to share a strategy that is different from those already reported, students with only one strategy are less likely to get a chance to report.

5. They understand that all mathematics is integrated with the culture of the classroom and with the world outside school. By helping their students encounter mathematics in familiar contexts, they enable their students to think about the problems in a natural way rather than be limited to a formal approach divorced from their experiences.

6. They realize that assessment is an integral part of instruction. As students reveal their thinking, the teacher has a better understanding of the ways in which the students think about problems and about the size of numbers the students are able to use. The teacher then uses this knowledge to make subsequent instructional decisions.

THE CGI PROFESSIONAL DEVELOPMENT PROGRAM

The nature of children's thinking that provides the basis for the CGI professional development workshops has been described by Carpenter, Carey, and Kouba (1990, pp. 111–31). This knowledge supplied the framework for everything else that followed, and nine hours of the twenty-seven-hour course were spent on it.

The workshop is designed to help the teachers understand how children solve a variety of addition, subtraction, multiplication, and division word problems that involve single-digit numbers, multidigit numbers, and fractions. The workshop also gives the teachers an opportunity to explore how they might use their knowledge in instruction. Teachers learned to recognize differences among word problems, to identify the solution strategies that children might use to solve different problems, and to organize these strategies into a hierarchy of levels of thinking. The workshop helps teachers structure the knowledge they have acquired about problem types and solution strategies.

In the same way that children's learning of mathematics in the classroom builds on their previous understanding, the activities of a CGI workshop are designed to build on the previous understanding of the teachers who participate.

The teachers discuss principles of instruction that might derive from the research and make plans for their own instruction that are based on those principles. Although instructional practices are not prescribed, the teachers discuss how they might use the knowledge they have acquired in assessing their own students' thinking and planning for instruction. Specific questions were identified for teachers to address in planning their instruction, but they were not told how they should resolve them. These questions included (*a*) How should instruction build on the informal and counting strategies that children use to solve simple word problems when they enter school? (*b*) Should specific strategies like "counting on" be explicitly taught? (*c*) How should symbols be linked to the informal knowledge of addition and subtraction that children exhibit in their solutions of word problems?

The teachers read articles (Fennema and Carpenter n.d.) that classify the problems by type, synthesize the results of research on children's solutions of word problems, and discuss how these findings might be applied in the classroom. Videotapes of children solving problems are used to illustrate solution strategies, and videotapes of teachers and students in classroom situations are discussed. A variety of instructional materials, including trade books, manipulatives, and enrichment materials, are also available for the teachers to review.

THE SCHOOL DISTRICT'S ROLE

The district mathematics supervisor and three first-grade teachers attended a six-day CGI workshop in February/March 1989. During the remainder of the 1988–89 academic year, the teachers explored how the knowledge they acquired from the workshops might be used in their classrooms. As a result of this exploration, the teachers made a commitment to use the CGI principles as the basis for their classroom instruction during the 1989–90 academic year.

Because of the enthusiasm of those three teachers, the district made a decision in November 1989 to make CGI workshops available to an expanded cadre of teachers during spring 1990. Because graduate credit was a stimulus to teacher participation, the workshop was offered as a course through the nearest campus in the University of Wisconsin system. The faculty for the course were members of the CGI project staff from the University of Wisconsin at Madison.

The course, which was held in the school district, duplicated the content and structure of the initial six-day workshop but was organized in three nine-hour sessions, each beginning Friday afternoon and ending Saturday afternoon. The district paid the teachers' tuition, and the twenty-four participants were awarded two graduate credits in education. The district mathematics supervisor and the building principals encouraged teachers to use the CGI philosophy and principles in their teaching. In spring 1991 this two-credit course was repeated for twenty-one additional primary-grade teachers. The course was repeated again in spring 1992. The original three teachers now feel confident that they could lead parts of the workshop but have not yet assumed responsibility for conducting the workshop for the district.

CHANGES IN TEACHING

The teachers entered the course with a variety of beliefs about how children learn. At the conclusion of the course all participants had changed, but they still varied in their understanding of children's thinking and in the ways they used the knowledge gained in the workshops. Interviews and classroom observations conducted in November 1991 revealed that a few teachers continued to believe that children cannot solve problems without direct instruction on facts and procedures. Some teachers expressed inconsistent beliefs. They reported that they were building on children's thinking and stressing problem solving, but their descriptions of their lessons indicated that they were not consistently attending to individual students' solutions. Other teachers appeared to be making good progress in the transition to an approach that reflects the CGI philosophy. Three of those teachers, who took the course at different times, are described next. Ms. R completed the course more than a year before the observations and interviews, Ms. J completed the course more than two years before, and Ms. M completed the course just a few months before. These teachers have responded to the course in somewhat different ways, just as children respond in different ways to mathematics instruction.

Teacher Change: Ms. R

Ms. R is a second-grade teacher with ten years of teaching experience, four years in second grade. "Before, I never really asked the students about their thinking. I told them what to do." Ms. R now sees her role as that of a facilitator. The students' role is to be actively involved and to think about the problems. She finds that "they all have some way of getting started. Story problems were hardly a part of math before. And that is the basis now." Before, children had trouble with story problems. "They were always difficult. You always knew there was going to be a bad day (when we were going to solve story problems)." Ms. R now believes that most of the children can be successful with most problems and that it is not necessary for her to explain how to solve the problems.

Ms. R notes that before CGI she had kids who would just sit and not engage in the mathematics. Now, even those who previously wouldn't do well with facts or practice pages are having success. "I am putting more of myself into it, and I am finding that the kids are putting more of themselves into it. They are enjoying it more. I think the lower-ability students are definitely much happier about math and can see that they can succeed in it. They might not have the way I taught before." She enjoys preparing for math class more now than before. "It is more creative, or it at least gives you a chance to be creative."

Ms. R now realizes the value of having children use manipulatives for problem solving. This is new. She also realizes that it is important for her to observe the students as they work on solving problems as a way of becoming familiar with the way each child thinks. "There is really a lot that is different because I think before we were looking at *the* right way to do things as opposed to a variety of ways."

Nevertheless, Ms. R seems to lack confidence that she is covering the topics of the second grade. She was more confident when she relied on the book. "If you follow the book, you feel that you are covering what you are supposed to be covering." She would like to be assured that she is doing the right thing. "I feel that what the students are learning now is more important than what they were learning before. I am a little uncertain about knowing if we are covering everything, but I feel good about the content I am teaching."

Her concern about covering the curriculum is also reflected in her concern about her students' performance on standardized tests. She would like assurance that her students are doing well. She does not have the students drill on facts but worries about their future. "I know that next year—unless things change—they will probably go into the math book in the third grade, so I do have some feeling that I need to do more, maybe get them used to a page of math facts."

Ms. R perceives the district mathematics supervisor to be very supportive of her involvement with CGI. She is less clear on how knowledgeable her principal is or how supportive he is. "I don't get a lot of feedback from him on it. I am not sure how informed he is. It probably would be good to sit down and talk to him. He has come in and observed." The previous year Ms. R had an opportunity to observe CGI teachers in another school. She would like to do more observing. It appears that Ms. R would gain more confidence if she could collaborate more with other teachers undergoing similar changes.

The CGI course made Ms. R eager to teach mathematics in a new way. She believes she now has a much better understanding of the ways in which children think about mathematics. In the workshop the teachers were encouraged to talk about the variety of ways in which children might think about a story problem. They saw videotapes of children solving problems. "We could see what the kids were doing. That was really much better than just reading it in the book." Contacts with experienced CGI teachers were also very important. The teachers' specific examples "mean more than reading research results."

Teacher Change: Ms. J

Ms. J is a first-grade teacher with eighteen years of teaching experience. She was one of the three teachers who participated in the initial CGI workshop two and one-half years ago, and she repeated the workshop when it was first offered as a course one and one-half years ago. In addition to her teaching responsibilities, Ms. J is the citywide elementary-level mathematics chair. As chair, she works closely with the district coordinator of mathematics and science, who was responsible for arranging for the district to participate in the initial CGI workshop.

Ms. J has changed the ways in which she thinks about students as individuals. "I used to think that all the students had to come up to a certain point, and I had to do everything I could to get them there. Now I think about it differently. I have to know how each child thinks and then help each one move to a higher level of problem type and a higher level of strategy." Her expectations of students have also changed. "I didn't give them enough credit for what they knew. They certainly know more and have more sophisticated strategies than I ever dreamed of. First graders are capable of learning so much more than I thought they could."

Partly as a result of these changes in expectations, her curriculum has changed its focus. "I did a lot of pushing the basic facts, and I spent a lot of time on speed tests. I don't do that anymore. We spend way too much time on drill and practice. If you can solve five problems, why solve a hundred?" Place value, which previously received no emphasis, is now an important component. "It was scarcely touched on before and it's so important to them now."

Her change in approach and emphasis has made her less dependent on the textbook. "I am so comfortable without a basal text now. I envisioned that being hard at the beginning, but the text just ties you down. It stifles you."

But these changes also create problems with transition as students leave CGI classrooms. "When our CGI children go into a non-CGI classroom, the parents complain that they're not challenged enough in math, which is true. They go back to reviewing basic facts and not much more than that. It is just not challenging and not fun." And she adds, "It's fun in CGI."

There was a mutual excitement in Ms. J's classroom that was new both for her and for the students. "When we came back from the workshop and tried some of those ideas, we saw our kids getting excited. Then we got more excited. I think the more we use it, the more confidence we gain." The students seemed to gain confidence also. "Math is wonderful for their self-esteem. That's one of the big pluses. The kids really gain a lot of self-confidence." Ms. J attributes the self-confidence to the children's success. "The better they do, the more successful they are and the better they feel about it. I've never heard any child whine or say, 'I don't want to do this' when it's math time. When I say, 'Get out your math bags,' the children say, 'Yea!' Sometimes they'll say, 'Can I stay in at recess and get it finished?' Some students used to say, 'I'm not very good in math.' I don't hear that anymore."

Ms. J has redefined her role as a teacher. She no longer thinks of teaching as explaining. "My role is not to demonstrate how to do the problem. It is to give the children a problem that helps them develop, that helps them move on to a higher level. I'm very careful not to do the teaching. I let the kids do the teaching."

Even though it has been three years since her first participation in a CGI workshop, Ms. J sees that her development as a teacher is gradual and continuous. "The first year it worked wonderfully. The next year it worked even better because I knew from my successes and failures what I didn't want to do." Continuing study is an important part of her development. "My understanding of CGI keeps changing. Every time I reread the articles, I see something I hadn't seen before. I'm getting better, but the only way I get better is to keep reviewing."

The district mathematics coordinator and the building principal have both been very supportive of Ms. J's development over the last several years. There are frequent requests for Ms. J to visit other schools in the district to work with teachers and to speak at mathematics conferences. "My principal is very supportive. I think he's very excited about CGI. He's a CGI advocate in the state now, too. He takes a lot of pride when representatives from other school districts want to visit our classrooms and are enthusiastic about what they see.

Teacher Change: Ms. M

In contrast to Ms. R and Ms. J, who have each had more than a year to grow in their ability to use their knowledge of students' thinking, Ms. M, who has been teaching first grade for twenty-seven years, has taught for only two and one-half months since completing the course. She is not yet able to recognize children's solution strategies and plan her instruction according to her knowledge of children's thinking. Nevertheless, she has made some changes that are important to note.

Ms. M described a lesson in which she had given each student a small bag of M&M's. Each student sorted his or her M&M's according to color and made a graph showing the number for each color. Then each row (students in Ms. M's class sit at desks arranged in rows) worked to find the total number of each color for the row. The numbers were in the hundreds. Ms. M was amazed that first-grade children were able to use such large numbers. "It's only November," she said. "Usually I am first starting addition at this time and adding maybe up to the fives or the sixes." After twenty-seven years of teaching, Ms. M has just discovered that her students already know a lot about counting. "Before, it seemed like a lot of them weren't interested in counting 105, 106, …."

She believes her students show more interest and self-confidence than they did before but is unsure why. She doesn't believe she is doing anything differently from what she did in previous years. But then she says, "Maybe I am going beyond what my expectations were. I was lucky to get students to add two numbers together and to write and just count the numbers, say, to forty."

Her awareness of the students' ability to work with larger numbers has created some problems for Ms. M. "Giving the children a freer range to go beyond is hard for me as a teacher. Students want to work with a number like 1199. I think they probably don't know what 1199 looks like, but they want to do problems like this. It's a little hard for me to let them just go on and do that; it really shouldn't be, but it is."

Ms. M has changed the emphasis she places on having students solve word problems. "I didn't use them at all, or very seldom. The only time I used them was when they were in the book. The children didn't like them. They were hard." But now her students are demonstrating abilities that astonish her. "It is unbelievable how they can come up with some of these answers! Really! I am amazed! And the way they can explain them to you. I can't believe it."

Ms. M attributes the children's success to her own higher expectations. "I think what has changed is the perspective I have on what the students can do. It still boggles my mind that they can do some of these problems that I would never have thought they could do. If you start them with word problems, your expectations are there and they will come through. And if you don't, nothing will happen."

In spite of the changes she has seen in her students, Ms. M still sees problem solving as subordinate to the completion of pages in the textbook; she has her students solve problems about twice a week. The district has adopted a new textbook this year, and Ms. M is unsure about pacing and "nervous about getting through the book." She doesn't want to abandon the textbook for a problem-solving approach because she would worry about not covering important concepts. Consequently, she plans to cover everything in the book.

Ms. M realizes that she is not using her knowledge of children's solution strategies as well as she might. "I am still trying to get my feet on the ground," she says. "If I had more time to just work on it by myself and do more thinking on it.… I need more confidence. Then I could give the children a freer rein. I can tell the other teachers are confident in just using CGI. They don't even use their textbook. I am not at that point. I am probably less confident after seeing all the ways to explain these problems. I am wondering, 'Wow! How am I ever going to learn all these ways?' I think that really kind of got me. It is going to take a while. I hope I can understand. It is still hard for me to understand how they can see it."

Ms. M recognizes the value of sharing experiences with other CGI teachers. There are two other CGI teachers in her building at her grade level and several others at other grade levels, but there has been no organized opportunity for collegial interactions and the teachers do not seem to get together spontaneously.

It is quite possible that Ms. M will continue to change, and she may yet become quite expert in interpreting children's thinking and responding to it in her instruction.

CONCLUSION

The changes in the ways in which these teachers think about teaching and learning may be attributed to one or more of the following features of the workshops in which they participated:

1. The workshop activities focused on research-based knowledge of children's thinking within specific mathematics domains.

2. An important workshop activity required the teachers to test the validity of the research-based knowledge by interviewing students in their own classrooms. This confirmed the claims made in the workshop about the variety of strategies children use to solve story problems.

3. Teachers were specifically asked to interview students who were having difficulty recalling facts. The teachers learned that the ability to recall facts is not prerequisite to solving story problems. Furthermore, they found that the ability to recall facts was a poor predictor of successful problem solvers.

4. The participation of experienced CGI teachers as part of the workshop staff was often identified as an important element in the success of the workshops. These teachers described ways they used CGI knowledge in their classrooms and attested to the effectiveness of the instructional approaches discussed in the workshop. The enthusiasm of these teachers made the workshop participants even more eager to experiment with these new ideas.

5. The workshop provided a structure to knowledge that was already known, in whole or in part, by the workshop participants regarding how children learn mathematics. Much of it did not conflict with what they already knew. However, because their previous knowledge frequently lacked structure, it was less useful to them than it was after it became structured in the workshop.

6. The ways in which CGI project staff interacted with teachers participating in the workshop paralleled, in many respects, the ways in which CGI teachers interact with their students. The staff recognized that the teachers started with a great deal of relevant knowledge, and they attempted to help the teachers build on that knowledge and structure it in ways that are similar to the ways in which CGI teachers attempt to help their own

students build on their existing knowledge. The CGI staff had a vision of how teachers might use their knowledge of children's thinking in their instruction, and they made their own instructional decisions in similar ways.

7. Interactions with students both during the workshop and after its conclusion were influential in shaping teachers' ideas about instruction. As Knapp and Peterson (in press) report,

> Teachers typically did not go into the CGI workshop already espousing interpretivist or constructivist principles. Rather, the actual experience of seeing their students generate solutions to complex mathematical questions led many of these teachers to a grounded understanding and belief in these principles which they had only begun to develop during the workshop.

Many innovative attempts to reform mathematics curriculum and instruction in the past failed partly because they attempted to place a new layer of knowledge on top of the teachers' existing knowledge and beliefs. When the new knowledge is fundamentally incompatible with the existing knowledge and beliefs, teachers tend to reject the new knowledge or modify it to make it fit their existing structures, much as children do. The CGI workshop is designed to help teachers modify their beliefs about teaching and learning. When teachers are equipped with these new beliefs and more structured knowledge, it is no surprise that dramatic changes in their classroom teaching often result.

REFERENCES

Carpenter, Thomas P., Deborah A. Carey, and Vicky L. Kouba. "A Problem-Solving Approach to the Operations." In *Mathematics for the Young Child,* edited by Joseph N. Payne, pp. 111–31. Reston, Va.: National Council of Teachers of Mathematics, 1990.

Carpenter, Thomas P., and Elizabeth Fennema. "Cognitively Guided Instruction: Building on the Knowledge of Students and Teachers." In *Researching Educational Reform: The Case of School Mathematics in the United States,* edited by Walter Secada, pp. 457–70. Special issue of *International Journal of Educational Research.* Elmsford, N.Y.: Pergamon Press, 1992.

Fennema, Elizabeth, and Thomas P. Carpenter. *Cognitively Guided Instruction: Readings.* Madison, Wis.: Wisconsin Center for Education Research, n.d.

Knapp, Nancy F., and Penelope L. Peterson. "Meanings and Practices: Teachers' Interpretations of 'CGI' after Four Years." *Journal for Research in Mathematics Education.* Forthcoming.

28

Professional Development through Action Research

L. Diane Miller
Neil P. Hunt

VARIOUS documents are calling for broadly based reform of mathematics education in curriculum development, assessment, and teaching. The most recent and possibly most widely publicized of these documents in the United States include *Everybody Counts* (National Research Council 1989), the *Curriculum and Evaluation Standards for School Mathematics* (National Council of Teachers of Mathematics [NCTM] 1989), and the *Professional Standards for Teaching Mathematics* (NCTM 1991). The *Curriculum and Evaluation Standards* created quite a stir in Australia. Soon after its release, the Australian Education Council released *A National Statement on Mathematics for Australian Schools* (1991). The *National Statement* is the result of a collaborative project by the Australian states and territories and the Commonwealth of Australia. According to the authors, it "documents areas of agreement between education systems about directions in school mathematics, the principles which should inform curriculum development and the extent and range of school mathematics" (p. 1). Like the *Standards,* its purpose is to provide a framework around which schools can build their mathematics curriculum, but it does not present a syllabus or curriculum for direct use.

Also like the *Curriculum and Evaluation Standards,* Australia's *National Statement* has implications for reform in how school mathematics is taught, which indirectly has implications for the professional development of mathematics teachers. Many Australian classrooms do not reflect the type of pedagogy and student activity that emphasize an understanding beyond getting the right answer. To implement the type of mathematics curriculum reflected in the *Curriculum and Evaluation Standards* and Australia's *National Statement,* practicing teachers must take a more active role in their own professional development.

This article was written by L. Diane Miller to capture the classroom experiences of Neil P. Hunt as a teacher-researcher.

A teacher can continue his or her professional development in various ways, including attending professional development activities sponsored by a school or local district; joining a professional organization like the National Council of Teachers of Mathematics or, in Australia, the Australian Association of Mathematics Teachers (AAMT); and reading publications like NCTM's *Teaching Children Mathematics, Mathematics Teacher,* and *Mathematics Teaching in the Middle School* and AAMT's *Australian Mathematics Teacher.* Another type of professional development activity for practicing teachers is action research. The term *action research* is a label for classroom investigations undertaken by teachers. It is a form of self-reflection that encourages teachers to be aware of their own practice, to be critical of that practice, and to be prepared to change it (McNiff 1988). The role of a practicing teacher undertaking action research is teacher-as-researcher. We hope that this article, which describes Neil Hunt's experience in an action-research project, will convince other teachers to consider the role of teacher-as-researcher as a meaningful way to continue their professional development.

A PERSONAL PERSPECTIVE

During 1990, I was introduced to the use of writing in mathematics classes to enhance the learning and teaching process. I had heard a university researcher talk about the idea and had read a couple of articles on the subject in professional journals. It seemed like a fairly simple approach to implement. I thought it had potential to be a rewarding, informal way to diagnose students' understanding of school mathematics.

The mathematics educators at a local university were organizing an action-research team consisting of schoolteachers and university researchers to investigate the benefits teachers derived from reading their students' writing in secondary school mathematics classes. Local teachers were invited to an afternoon meeting at which the university researchers talked about the results of research on using writing in mathematics classes, the design of the proposed study, and the responsibilities of each person involved. The most important point made that afternoon was that the teachers had to feel ownership of the study. From the perspective of the university researchers, the study would not be successful if the teachers participated for the benefit of the university researchers. The intention was that the teachers participate because they want to investigate an alternative approach to learning how well their students understand the mathematics being studied in school. I was one of three teachers from three different schools who volunteered for the project. With very little knowledge of, and no experience in, conducting classroom research, I assumed the role of teacher-as-researcher.

THE PLAN

I chose a Year-11, introductory calculus class in which to implement a writing activity. I decided to allow the students to respond for five minutes in approximately three out of every five instructional periods to a writing prompt that I had composed. Writing prompts are simply worded statements or questions directing students' thoughts to the explanation of a single concept, skill, or generalization (Miller 1991). Prompts can also encourage students to express their attitudes and anxieties about mathematics or problems they encounter while learning mathematics.

One way to present the prompt is to reproduce it on a piece of paper and allow the students to write on and return that paper. Some teachers ask students to keep a journal in which they copy the prompt from the chalkboard or an overhead projection and then produce their response. These responses can be collected by the teacher after every writing or at the end of a week. Copying the prompt takes students' time from thinking, writing, and revising. If, however, paper is scarce or access to reproduction equipment is limited, copying is a viable alternative. Since my resources were limited, I projected the prompt or wrote it on the chalkboard, but I did not ask the students to copy it. I kept a copy of the prompt with each set of papers. I used my wristwatch to monitor the time; however, other teachers use a kitchen timer.

Because this action-research study focused on the benefits teachers derive from reading their students' writings, the three teacher-researchers kept individual journals in which we wrote our dominant impressions about what we were learning from the students' writings. This task was perhaps the most difficult. Like our students, we were not accustomed to writing about what we were learning in mathematics class.

The team of three teacher-researchers and two university mathematics educators met twice during the first ten-week term of the 1991 school year to discuss the design and implementation of the study. We decided to implement the writing activities during the second and third ten-week terms. The team met once during each of these terms and again at the end of the study.

IMPLEMENTATION

At first I was anxious about composing the prompts. To my relief, the focus of a prompt seemed to evolve naturally. I generally directed the students' attention to a topic being studied or a previous topic they needed to review in preparation for a lesson. The following is the first prompt I gave students:

> Your friend is still having difficulties with some concepts involving sequences. Try to explain clearly to your friend how to use the relationship
>
> $$t_n = \frac{1}{2} t_{n-1},$$
>
> where $t_2 = 6$, to derive the first four terms of the sequence.

Two of the seventeen students in the class had no idea about how to respond. Half the group could do the problem, but their reasoning contained incorrect statements. They followed a correct procedure to find the first four terms but had difficulty with the explanation. A comment in my journal made after I read this first set of writings reminded me to do some follow-up work, perhaps reteaching, to clarify the students' understanding of recursive sequences.

Two days later, I gave them a second prompt.

> Imagine that you are writing a note to your best friend to explain how to differentiate a polynomial. Assume that your friend really wants to learn how to differentiate a polynomial and that she or he must rely entirely on your explanation for help.

This prompt asked students to recall a process that had been studied three weeks previously. Most explanations centered on what to do and how to apply the rule for the derivative of cx^n. Most students' recall fell far below what I thought were acceptable levels. As in the writings produced for the first prompt, I found many correct procedures along with mathematical errors of different types. A comment in my journal on this day was, "At this stage I'm already concerned that a number of students do not have a good understanding of key mathematical words."

To assist in my own analysis of the students' writings, I designed the scoring system in table 28.1 to represent my opinion of their writings. After reading a set of writings, I could easily produce a total score for the class. This score sometimes suggested that I reteach a lesson with the whole class or speak to students individually about their understanding of the topics being covered.

Table 28.1

Score	Explanation
2	Explains correctly; reflects understanding
1	Has difficulty explaining, but can work the problem (when applicable)
0	Cannot explain (leaves blank) or gives an incorrect explanation

After five prompts, I made the following general comments in my journal:

- "The worst responses are definitely coming from students whom I perceive to be the weakest in the class."
- "With a test coming up in a week's time, I will certainly review the topics covered in prompts 3–5. The students have very little, if any, real understanding of those concepts."
- "My first prompt next week will ask students to list specific concepts that they are having difficulty with. Perhaps they will suggest topics to review that I haven't thought about."

LESSONS I LEARNED AS THE PROJECT PROGRESSED

Students are not accustomed to writing in mathematics classes. These students were above average—probably some of the best mathematics students in the school. However, they initially struggled in their attempts to express in writing their understanding of the mathematics being studied in class. As time passed and more prompts were used, their writing improved in two ways. First, they started writing more in the five-minute period, and second, their written explanations improved.

Initially, the class complained about having to write in mathematics class. A few students said they would "rather do mathematics than write." They obviously did not perceive writing about their understanding of mathematics concepts, skills, and generalizations as "doing mathematics." As time passed, their complaints subsided and we settled into a comfortable routine, perhaps because they knew I was more committed to the task than they were to changing their attitudes about the writing.

Some students, however, did not like the writing prompts at all. I suspect that they disliked the activity because the writing allowed me to diagnose more accurately what they did or did not know. Some of them were fairly accomplished at manipulating symbols and following an algorithm, but they resented the writing activities because the writing prompts forced them to demonstrate a different type of understanding that they did not have.

Very early in the study, I became concerned about the students' misuse or lack of use of mathematics vocabulary. This concern persisted. The following was my fourteenth prompt:

Use each of the following terms—*index, coefficient, numerator, denominator, fraction, function,* and *variable*—in an explanation of why $f(x) \neq g(x)$ where

$$f(x) = \frac{4x^{-1}}{3^{-1}}$$

and

$$g(x) = \frac{3}{4^x}.$$

The students' performance on this prompt was very disappointing. In most instances, the mathematical reasoning behind the students' assumptions was incorrect. The term *coefficient* was consistently used incorrectly. A comment in my journal reminded me to review negative indexes and use more prompts that focused the students' writing on the correct use of mathematics vocabulary.

At the first team meeting after the students began writing, I shared my frustration about the students' nonuse or misuse of vocabulary. The other two teachers involved said that they, too, were noticing the same problem in their students' writings. We asked the university team members what we should do. They reversed the question and asked, "What are you going to do?"

When I returned to my class the next day, I was more attuned to my students' use of appropriate vocabulary in their oral discourse. I was surprised to discover that they also avoided the use of technical language in their verbal comments. For example, they would talk about "the number in front of *x*" rather than use the word *coefficient*. They knew what *coefficient* meant. Why didn't they use it in their oral and written discourse?

Within a few days, one of the university team members came to visit my class. We were discussing various functions and how their graphs are affected when coefficients and constants in the functions change. She was taking notes as if she were a student in the class. At the end of the class, she asked, "Were you aware that not once in the whole lesson did you use the word coefficient?" At first I didn't believe her. Suddenly I realized that she was right. I had repeatedly asked the students to tell me what happened when we increased or decreased "the number in front of *x*." Why had I not used the word *coefficient?* She also indicated that I had said *the answer* rather than *the quotient* and that several times I had referred to *the number underneath* rather than *the denominator*.

After that time, I made a deliberate effort to use mathematics vocabulary in my instructional comments and encouraged students to do the same when talking to me or to one another. Sometimes I would use a specialized term and then stop to ask a student what the word meant. I always encouraged students to verbalize the meaning in their own words and not necessarily repeat a textbook definition.

Mathematics teaching and learning is an interactive process that depends on the understanding of carefully defined terms and symbols. Teachers must change their classroom discourse to use the correct vocabulary. We must become better role models for our students. What I learned by using writing in that introductory calculus class now influences my teaching practices in my other classes. I am consciously using the vocabulary and insisting that students use it more in their oral and written work. I previously had intentionally used more common language with my lower-school students (Year 8) hoping to demystify mathematics. I now believe that doing so to the extent of using specialized mathematics vocabulary infrequently is a mistake.

I am not suggesting that we go back to the days when students were held accountable for knowing the difference between *number* and *numeral* and were chastised for their misuse. Rather, I am suggesting that teachers use mathematics vocabulary within a meaningful context and encourage students to do the same. During this investigation, I learned that after I started using proper mathematics vocabulary in class, the students started becoming more comfortable with the language and the words automatically came to their mind a lot more quickly in their oral and written comments.

CONCLUSION

Undertaking professional development through action research can be very rewarding and enjoyable. Although preparing the prompts, reading the students' responses and writing in my own journal took time, the learning experience was worth the time invested.

The writing experience was beneficial to the students because it gave them an opportunity to express their understanding of mathematics in writing, which included the appropriate use of mathematics vocabulary. Being able to "express mathematical ideas orally and in writing" is an objective of the NCTM's *Curriculum and Evaluation Standards* (1989, p. 140). *A National Statement on Mathematics for Australian Schools* (1991) also stresses the importance of students' being able to communicate mathematically. The authors of the *National Statement* say, "A command of mathematical terminology ... is essential in learning mathematics and is part of numeracy" (p. 13). Thus, my action-research experience has influenced me to change my teaching practices in a way that is more closely aligned with reforms in mathematics teaching in Australia and the United States.

I benefited in other, more general ways, from reading my students' writings. For example, when their writings reflected complete lack of understanding or misunderstanding of a concept, I immediately retaught the lesson. On occasion, I scheduled a revision based on what was learned from the students' writings about topics taught some days or weeks prior to the writing prompt. I was also able to initiate private discussions with individual students whose writings indicated special needs. Finally, the writings were beneficial to me because they afforded each student the opportunity to communicate with me whenever a prompt was used. Even though the class was small—seventeen students—I sometimes was unable to have a personal communication with each student on the days they did not write.

My role in this project was teacher-as-researcher. Perhaps a better title is teacher-as-learner. Research should be viewed as a learning experience and a responsibility of all teachers who seek to improve their teaching and learning. Attending professional development days organized by the school, reading articles related to teaching, joining a professional organization and attending meetings sponsored by the organization are important and can be beneficial professional development activities. However, in my opinion, professional development through action research offers rewards that can affect teachers' practice for the remainder of their career.

REFERENCES

Australian Education Council. *A National Statement on Mathematics for Australian Schools*. Carlton, Victoria, Australia: Curriculum Corp., 1991.

McNiff, Jean. *Action Research: Principles and Practice.* London: Macmillan Education, 1988.

Miller, L. Diane. "Writing to Learn Mathematics." *Mathematics Teacher* 84 (October 1991):, 516–21.

National Council of Teachers of Mathematics. *Curriculum and Evaluation Standards for School Mathematics.* Reston, Va.: The Council, 1989.

———. *Professional Standards for Teaching Mathematics.* Reston, Va.: The Council, 1991.

National Research Council, Mathematical Sciences Education Board. *Everybody Counts: A Report to the National on the Future of Mathematics Education.* Washington, D.C.: National Academy Press, 1989.

29

Stop, Look, Listen: Building Reflection into Continuing Professional Development

Jane F. Schielack
Dinah Chancellor

$\mathbf{F}$OR THE past three years we had been conducting extended projects in mathematics instruction for our district's teachers in kindergarten through grade 8, building familiarity with the *Curriculum and Evaluation Standards for School Mathematics* (National Council of Teachers of Mathematics [NCTM] 1989), the *Professional Standards for Teaching Mathematics* (NCTM 1991), and instructional approaches such as cooperative problem solving and mathematics journal writing. Suddenly, we were faced with a group of teachers who were not interested in being "sold" on philosophies or techniques; they just wanted to know what to do! No matter how many activities we tried to squeeze into our three-hour, semimonthly sessions, we left with cries of "More, more!" ringing in our ears .

Yet we felt dissatisfied with the group's focus on accumulating activities. Their journal entries discussing the implementation of these activities in the classroom lacked depth and purpose. In analyzing our discomfort, we became aware that we needed to do with our teachers exactly what we constantly encouraged them to do with their students—provide time, and structure, for productive reflective thinking.

TIME TO STOP

Our first decision was to allow thirty to forty-five minutes at the end of each three-hour session for teachers to reflect on the activities introduced during the session. This period of reflection would be used by the participants to focus on

The activity that is the subject of this article was produced under a grant from the Texas Higher Education Coordinating Board and the U.S. Department of Education under the auspices of the Eisenhower Mathematics and Science Grants Program, Title II. Opinions, findings, and conclusions expressed herein do not necessarily reflect the position or policy of the Texas Higher Education Coordinating Board or the U.S. Department of Education, and no official endorsement should be inferred.

the objectives of the activities, the types of mathematical thinking that students should be engaging in as they experience the activities, and the decision making in which the teacher would be involved before, during, and after the activities. Their decision making would involve choices of materials, management techniques, motivating introductions, questioning strategies to encourage student exploration and generalization, and appropriate means of assessment.

The "stopping" time for reflection would be a serious encroachment on our previous activity time and would therefore be viewed very critically. In the teachers' eyes, the loss of quantity would have to be offset by an observable gain. We would need to offer a supportive environment until they began, as individuals, to reap the benefits of a reflective attitude toward their teaching.

TIME TO LOOK AND LISTEN

To supply a structured setting that would encourage productive reflection, we designed the outline in figure 29.1 for the teachers to use as a guide for thinking about and discussing each activity. The purpose of the outline was to emphasize the teacher-oriented aspects of an activity that have an impact on its effectiveness. What is the mathematical purpose (or objective) for doing this activity? What concepts will be developed or strengthened? What decisions need to be made about grouping, resources, and time? How can it best be introduced? What could be done or said to encourage student exploration? What guidance should be given to help students summarize and generalize their understandings? What questions, observations, and tasks can be used during and after instruction to assess student learning?

Although the teachers used the outline to guide their reflective thinking at the end of each activity session, they also began to refer to topics on the outline during the session. In anticipation of their discussions during the reflection period, the teachers began paying more attention (*a*) to the questioning strategies we modeled as we did activities, (*b*) to the mathematical goals of the activities, and (*c*) to the things being done and said by the other members of their groups.

REAPING THE BENEFITS OF REFLECTION

The final step in exemplifying the worth of reflective thinking was the generation of a useful product. During the reflection period, the teachers were encouraged to make notes of their discussion on their outline forms. Each group was asked to turn in one outline showing the ideas generated by the interaction of the members. We then compiled the groups' ideas into a single class outline of the activity to be distributed at the next meeting. The final outline did not include every suggestion turned in but rather a selection of the best ideas in each section, the result being better than any single group's product, yet owned by all members of all groups.

TITLE OF ACTIVITY

Objective:

Math Concepts: (List)

**Concepts from
Content Areas:** (List)

Materials: (List)

**Management
Suggestions:** Prerequisite skills, things to prepare ahead of time, student groupings, time frame

Procedure:

Introduction: Specific suggestions for introducing the lesson

Explorations: Questions to help students while they are working

Extensions: Specific math ideas for those groups who finish early

Summary: Questions to help students reflect on learned mathematics

Student Assessment

Questions: *To help uncover what students have learned from the activity* (could be part of the summary discussion) (*Example: In a primary-level activity involving trains of 10s and extras*—"Explain your strategy in predicting whose train was longer. How many more cubes did you have than your partner? How do you know? Explain how you decided which number to record in the tens column. The ones column. What number did you make?" *In an intermediate-level activity using the 1, 0, +, ×, and = keys on the calculator to investigate different ways to use powers of 10 to display a telephone number*—"How did you use place value in designing your approach? How did you organize your procedure so that everyone understood it? Would it have been helpful to have had the use of other keys also? Can you make a general statement about how to generate any given phone number?")

Observations: *Evidence of understanding to watch for during the Exploration or Summary, such as specific behaviors, explanations, or answers* (*Example: In a primary-level activity involving trains of 10s and extras*—Were students able to break their train into groups of tens and extras? Did they accurately record their number of tens and ones? Could they read their number? Could they explain the relationship between the digits in the numeral and the cubes? *In an intermediate-level activity using the 1, 0, +, ×, and = keys on the calculator to investigate different ways to use powers of 10 to display a telephone number*—Did the students' explanations reveal an understanding of place value? Were students able to explain clearly the use of powers of 10 in accomplishing their task?)

Tasks: Journals, written summary statements, and so forth

Fig. 29.1 Outline form for teacher reflection during professional development activities

The teachers' reactions to the final product were exciting and encouraging. They recognized their own contributions and realized how their ideas had been enhanced when joined with others. To verify the importance and usefulness of their reflective thinking, we announced that from now on there would be no handouts from us to describe the activities we did. The standard handouts would be replaced with the outlines that the teachers would generate at the end of each session.

CONCLUSION

The change in the atmosphere of the activity sessions was immediate, from "What are we going to do next?" to "What other questions could I ask in this activity? What task would be most appropriate for determining if students understood the idea?" The compiled activity outlines generated such enthusiasm that we expanded from outlining one activity each session to three activities each session by randomly assigning an activity to each group or by breaking into grade-level groups and assigning one activity for each grade level, with a minimum of three groups working on each activity.

The teachers' growth in thinking reflectively about their teaching has been evidenced not only in their group participation but also in the number of analytical comments appearing in their journal entries and in the quality of the questions being asked in their classrooms. What we have learned is that for our teachers, just as with our younger students, "less can be more" if they take responsibility for their own professional development and become actively involved as producers instead of just consumers.

REFERENCES

National Council of Teachers of Mathematics. *Curriculum and Evaluation Standards for School Mathematics.* Reston, Va.: The Council, 1989.

———. *Professional Standards for Teaching Mathematics.* Reston, Va.: The Council, 1991.